AN UNCHARTED PROGRESSION

An Autobiography by Robert Lee Howard

Fulton Books
Meadville, PA

Published by Fulton Books 2023

ISBN 979-8-88505-237-5 (paperback)
ISBN 979-8-88505-238-2 (digital)

Printed in the United States of America

Preface

Abraham Lincoln or Honest Abe, as he has often been called, once said, "The best way to predict your future is to create it." Malcom X, a Muslim leader, was also an unapologetic promoter of the *by any means necessary philosophy*. Some may regard this as an odd way to begin what is essentially an autobiography. How could anyone effectively express their viewpoints while highlighting two vastly different individuals? Malcolm X was a self-admitted robber, womanizer, drug user, and ex-convict. But he was also a man who refused to be silent about the sexual abuse and victimization of faith-bound women by their Muslim leader. Lincoln is closely associated with the September 22, 1862, Emancipation Proclamation. That nation altering executive order, that freed the slaves, became enforceable on January 1, 1863. But the proclamation's intent was not to integrate slaves into the *melting pot of* American culture. It was in fact hoped that the slaves would reestablish themselves outside of the United States. The delay in the proclamation's enforcement was to give states an oppurtinity to rejoin the Union. Still, Lincoln realized that slavery had to be abolished in its entirety to ensure that it would never again be a nation dividing issue. We, the recipients of these questionable actions, must realize that leaders, worthy of hero status, must have more than fine minds. Our leaders in any profession, must have the moral integrity to avoid creating or approving enforceable directives that promote division, inequality, fear, misrepresentation, lies, bigotry, misdeeds, and cover-ups. Ideally, non-self-serving leaders will only create and endorse necessary directives that will be applied equally to all.

Many leaders will support any issues, right or wrong, if their careers are threatened. But Lincoln and Malcolm X, put the needs of others before their own safety, even though those actions led to their assassinations. Personally, I was once so clueless that it is a miracle I am still here, revealing my risky paths. What could a person of color, entering this world with nationwide imposed limitations, be expected to achieve? Yet, others before and after me rose above extremely difficult circumstances, including the denial of educational opportunity. I foolishly shunned much needed education early in life and, significantly reduced *the early inward motivation and drive* children need to build character and competence. Only my accidental exposure to opportunities which normally would not have been available, enabled me to become aware of and attempt to do something about my own short sightedness.

I wrote this book because of my concerns regarding *Covid-19, social issues,* and emerging catastrophes in the form of deforestation, pollution, *global warming, and hate.* Today hate is widespread and the many political and violent activists, embracing the *by any means necessary* patterns of behavior, must come to understand that *there are no wrong ways to do the right thing.* The 13th Amendment to the U. S. Constitution, ratified in 1865, abolished slavery. However, the fifteenth amendment to our constitution regarding the rule of one man one vote is being widely suppressed by some governmental officials to discourage voting. We cannot force some people to do the right things, but we can distribute vital information. Voters must be made aware of the lawful procedures, so that their ballots will not be discarded. They must then vote, in a timely manner, according to their own fact supported convictions. Citizens should not remain voiceless and stand passively by while letting outwardly vocal distorters of the truth influence the masses in ways that do more to create hate and division rather than love and unity.

I understand I am an author without a platform. But if I can get enough people to join with me in promoting earth saving actions, then our collective voices can grow and be heard worldwide.

Robert Lee Howard

This book is dedicated to my mother, Rose Lee Ayers-Winfrey-Howard. She always placed the needs of her family before her own needs. She will eternally have this son's respect, devotion, and special love.

Rose Lee Ayers

Siblings: Marie, Robert, Ruth, Oscar, Eileen, Frank, & Louise

Chapter 1
My Ancestors and Me

My mother, Rose Lee Ayers, was the daughter of Priscilla Ayers, who reportedly was part Native American. Wiley Ayers, my grandfather, grew crops that he sold to support his family. The land was in a rural farming community on the outskirts of Huntsville, Alabama. There were many other Ayers families living nearby. Rose's siblings included Datie, Donna Lee, Emma, Wiley Jr., Fannie, Roxanne, Simon, and Solomon. They all worked the land with little rest, so, to the surprise of no one, Rose left and went to work as a housemaid in Chattanooga, Tennessee. She found love in Tennessee, and gave birth to Artist, her first child, before she was twenty years of age. Evidently, birth control was a term neither she nor Solomon Winfrey, her first husband, ever thought about. Still, they were saddened when two of their babies were stillborn. Death visited our family again when Cornelia lived to be only six years of age. The effect of losing a named child was evident when my mother named her very next female child Cornelia as if, I think, by doing so she could somehow bring the first Cornelia back to life.

Rose and Solomon's search for a better life led them to leave Tennessee and travel steadily north, finally settling in Newark, New Jersey. However, life remained harsh, especially for those with limited job skills. The many frustrating years of hard work added up and gradually took a toll upon Solomon's state of mind. As the years went by with little change, he turned more and more to alcohol to reduce his conscious awareness of the endless stresses that he felt. He, therefore, unsurprisingly, developed alcohol related health problems that led to his death. When he died my mother, pregnant again with her seventh surviving child, Marie, lacked the help of any support group. It was then that she and her children moved to Bayonne, a small and overcrowded city, that was across the bay from Newark. In Bayonne, she was hired to cook for a boarding house whose tenants included Frank Howard, my father.

My father never talked about his family, but we learned through census and DNA records, that he was born in Columbus, Georgia in 1898. His brother, Willie, was born in 1887. George Howard, their father, was born in 1866 and lived in Clay City, Georgia. George's father, Leonard Howard, was born in 1842. He and his wife Julia of Bluffton, Georgia, were my great great-grandparents. They were also a connection to that time when my ancestors were slaves and worked as farmers.

Americans annually celebrate a creation of German folklore, called Punxsutawney Phil on February second. He resembles a large squirrel and is generally referred to as a ground hog or pig. That date is also the midpoint between the December winter solstice, when one of Earth's poles is at its' full tilt from the sun, and the period of March 19-22 in the northern hemisphere, when days are longer than the nights. If the groundhog upon first leaving its hibernating den sees its shadow, that would be a sign for it to return to its den to avoid six more weeks of harsh winter weather. Finally, it is a date in 1930 when I, Robert L Howard, the twelfth of fourteen children, was born. Kids hoping to avoid more bad weather teased me every year with chants like *don't let the sun see you spying piggy; go back into your den!*

Chapter 2
Our Homes

My first home on Fifty-fifth Street was one of many that housed about one hundred thousand people crowded into Bayonne. The three-mile-long city had one ninth grade junior high school, and one tenth, eleventh, and twelfth grade senior high school. Several elementary schools were scattered around the city. All the Schools were fully integrated and people, from many different nations, interacted with each other and got along surprisingly well. Our next home was a third-floor apartment on Nineteenth Street. It had two bedrooms that were sandwiched between the kitchen and the living room. The toilet in the hallway was shared with the family next door. The living room doubled as a third bedroom and had one window. It provided a view of the sidewalk and the steps as you entered the building. Our railroad apartment was so named because to get from the kitchen to the living room, you had to pass through two windowless bedrooms. Personal privacy did not exist. The kitchen had a wood burning, pot-belly metal stove that doubled as a heater during the cold winters. There was a sink with cold running water. Also taking up space was a hand me down electric washing machine. After depositing water that had been heated on the stove into the washing machine, it had to be rolled to the sink where a water hose was used to add cold water. The machine was emptied by using a pot to repeatedly scoop out the water. A rotating hand operated clothes rinsing device was attached to the machine. A big, round, corrugated tub was used to scrub heavily soiled clothing. The tub doubled as the children's bathtub. Eleven members of our family, including our mother and father, stayed in that tiny dwelling. Frank Jr. my mother's last child was about to be born. He and I slept with our parents after his birth. Artist, my mother's first child, did not live with us. The October 29, 1929, collapse of the stock market occurred just four months before I was born. My parents, thankfully, continued to work.

A tenant on the second floor was an old man who might have been in his fifties. He wore dark baggy clothes, but his overly friendly manner enabled kids to feel safe around him. He would often be heard to say, "Don't say it unless you mean it" or, "There is no wrong way to do the right thing." He was a cult follower whose chosen leader was an individual he called Father Divine. We often heard him singing about how Father Divine sure was good to him. Most of the kids made fun of Mr. Amos but I, at three years of age, tried to listen to him and become a better person.

I don't know how I, while still a three-year-old, could have such a clear understanding of an important family event. I remember standing between my father's legs as we looked down at the street from our living room window. We had only been there a short time when Oscar, my brother, appeared below. As he was about to enter the building he looked up. It was then that my father in an unspoken movement of his hand told his stepson, to stop. With another wave of his hand, he directed Oscar to keep moving. Oscar never lived with the family again. Oscar, who usually supervised us was gone. Because our parents worked, my younger siblings and I were left unsupervised. The evictions of other first marriage siblings continued until only two girls, Marie, and Estella, continued to live in our home. The building, as expected, was also condemned and we moved into an Eighteenth Street, second floor, two-bedroom apartment behind a three-story building. A six feet wooden fence was on the right and two outside storm cellar doors were on the left. The doors led to two storage areas under the first-floor apartment. Cracks in the floor of the lower apartment, allowed people to see into our storage area when its light was turned on. This house had hot and cold running water and was equipped with an interior toilet and a bathtub as well.

Shortly after moving, I was in danger of being added to the list of children who died young. Spinal meningitis would have caused my death. One of the causes of the disease that can affect the membrane of the brain and/or the spinal cord is bacteria. Without timely and effective treatment, the mortality rate in some cases might be as high as 70%. However, some old wives tales say that survivors of the disease are doomed to become blind, crippled, and crazy. After I was released from the hospital, my mother massaged my legs daily. It was a display of the love she felt for all her children. But some of my siblings were not sympathetic. They did not realize or care that I had been at death's door. My brother Eddie laughed as he told me, "You can see where you're walking, so just being crazy isn't the worst thing that could have happened to you.

Chapter 3
Environmental Influences

Tommy Lee lived with his father on the first floor. He was an example of everything I thought a hero should be. He was big, strong and, as far as I was concerned, he knew everything. Years later, he did not seem to mind when I found ways to attach myself to him. I think he liked my sister Eileen. Tommy had already formed a strong bond with Sonny, his friend who lived around the corner. It did not take long before we appeared to be the three musketeers, always doing things together. What made this relationship so special is that Tommy, a natural-born leader, allowed Sonny and me to sometimes take the lead when choosing things to do. So, I grew up learning how to be both a leader as well as a follower.

The lower level of the three-story building in front of our house, was occupied by a pool hall. Its owner and most of its customers spoke Spanish. The second floor was home to an Italian woman with seven children. My later school association with Raymond, one of her boys, enabled me to bond with the rest of the family. I loved the mother's spaghetti and sometimes when she called her children to dinner, she called me as well. The thing that was so special, about what is still a great meal for me, is that she served spaghetti with tasty, bread softened, garlic, and onion flavored meatballs. Two young Italian sisters occupied the apartment on the third floor. They were nice, and I sometimes ran errands for them.

I started a fight when I was in the first grade with a classmate named Willie Tomasevic. He kept loudly and repeatedly stating that Russia was the greatest country in the whole world. It wasn't a matter of politics, a word I'd never heard, but when I hit him all he did was stand there and cry. I felt bad about what I had done, but I could not bring myself to tell him I was sorry.

My father worked hard five days a week and partied so much harder on weekends. Men gathered in Mr. Lee's first floor bachelor apartment. They drank heavily, gambled, and sometimes fought each other. While hanging around my father, I learned the words, without understanding the meaning, to their bawdy songs. I tried to sing one of those songs, to my pretty third-floor Italian neighbor. However, I couldn't understand, at the time, why her face turned red and lost its smile. She then silently turned away, went back inside her home, and closed the door.

The first-floor man's den was not only a magnet for men; it attracted women as well. My father's obsession with his weekend retreat, led to domestic arguments. He was better at getting drunk than he was at gambling and his losses often led to the non-payment of bills. I heard my mother say she'd rather have a drinking man than a gambler many times. Ruth, my youngest sister, said she jumped on our father's back while hollering, "No Daddy no!"; to keep him from beating our mother. The *Might Makes Right Theory* only proves that the batterer is physically stronger and lacks the mental skills to sway others to their beliefs. While the women I have dated are not convinced that I am always right, I have never used physical violence to make them agree with me.

School was not my favorite destination. Our vice principal, Miss McInerny's visits to my first-grade class didn't help matters. When she came to my desk, she looked at my paper and said, "You write back-handed". My questioning expression that begged her for an explanation, was met with silence as she moved on without further comment to the next student. I guess she figured I couldn't understand. When my teacher kept me back, I told myself it was because she liked me more than all the other kids that she sent to the second grade. When I finally got to the second grade, I liked my teacher, Miss Maloney. She dated and married my P. E. teacher, Mr. Siegel. Physical education was the one class in which I did well. She visited me in the hospital after my tonsils had been removed, and the ice cream she gave me was delicious. Even though I can't recall anything that I learned, I was promoted to the third grade.

Tommy Lee and Sonny were ahead of me in school, so I established friendships with some of my White classmates. I spent a lot of time in Richie's house playing with toys that my family could not afford. Sammy, a Russian friend, used to delight in goading Wilma D, a woman-sized child, into causing our rather small teacher to cry and then laughed so loud that we had to shut the windows.

Chapter 4
Pedophile Prey

With my parents working and the older siblings evicted, the younger children were unsupervised. This was a period in my life when I had never thought about sex. So, when two brothers had me along with other neighborhood boys engage in sexual activities, we didn't even consider whether what they were asking us to do was right or wrong. The thing that puzzles me is that I, a know it-all according to my mother, should have known better. That changed when an abuser brought a much older guy under the porch to observe us and then offered to have me perform a sexual act on his friend. I don't know why, but suddenly I refused. Maybe it was the fact that I was in the same early elementary grade as this guy's much younger stepbrother. As if by some suddenly awakened insight, it dawned on me that what they were asking us kids to do was wrong. I refused to perform a sex act on him and went home. Days later, someone had done a lot of talking in the neighborhood about things I had done. I feared I would lose my friends. However, Tommy Lee, Sonny, and my sister Eileen, sat down with me, in the dirt covered yard. I listened to their advice even though I had already decided I would not be victimized again.

A few days later, while playing hide and go seek, I hid between my house and the fence. It seemed like a good place until Johnny Bronowski, who lived in the building with my molesters, appeared and forced himself on me. He only stopped when another child was heard coming. Playing any game after being physically and sexually abused was the last thing on my mind. I wanted to hurt somebody. I visited my friend Richie and left with his plastic toy bat. I then waited on the porch of my abusers until one of them came outside. I imagine the last thing he expected to see was an angry little kid with a bat. He stumbled and fell as he tried to avoid the bat, but I hit him and swung the bat again. As I struck the second blow, the bat flew out of my hands. I did not stop to pick it up. I ran blindly across the street, rushed into my room, and hid under the bed; I was afraid the cops were going to come and take me to jail. I did not tell my parents about being abused and after a day of not seeing any police, I was pleased with myself. I had solved the problem all by myself…or so I thought. The full truth was that two adult brothers caught their younger brother sexually abusing his own little sister. They also learned there were others who had tried to molest her as well. The brothers went on a rampage and, without calling the cops, battered all the people who were even suspected of being child abusers. Their punishing campaign was effective for me because I was never abused again. But I was proud because I had bravely attempted to fight my own battle. Years later I read about abuse victims, who killed themselves. While I may still have some lingering concerns, I never felt personally at fault about what, in realty, were crimes by pedophiles.

Chapter 5
Household Responsibilities

My mother scheduled chores for all her children and the one I hated most was washing dishes. Stella agreed to wash my dishes and I paid her with the money I earned by shining shoes and running errands. Momma was also the family doctor and gave us either cod liver oil or castor oil mixed with orange juice as cures. Finally, I didn't fear coughing around my mother because Stella secretly swallowed the medicine meant for me. But I had taken castor oil and orange juice so many times before that I could not erase its horrible taste from my memory. I was well into my forties before I could drink plain orange juice without it also tasting like castor oil. My brother Eddie was a teenager when I was born, and I defied him. Whereas Stella might regard me as funny, his response, was to brush me aside. I repeatedly told him, "When I grow up to the wall, I'm going to beat you up".

My mother worked hard six days a week, but on Sundays she became a Baptist through and through. She encouraged each of us to accept and worship our Lord and Savior Jesus Christ. In fact, Frank and I were baptized—I'm talking about being totally submerged—by Reverend Watkins, on the same day in Friendship Baptist Church. I was eleven and Frank was eight. I even sang in the choir. After church she read her King James Version of the Bible to us. I can't remember my father ever attending church. Ernestine V, a second-grade classmate, lost my hat. When my father discovered the loss, he beat me every morning, before going to work. Mercifully, the weekend arrived and the two days of drowning himself in whiskey dulled his recall. Monday morning found him wondering why he had been beating me and I was not about to refresh his memory. Please do not ask me about my father's sense of humor. He didn't have one. I once asked hm if he wanted to see a monkey. When he said yes, I held a mirror up to his face and got another beating. Still, because I was older than Frank Jr, he took me with him, I guess for company, when he went fishing.

The Bayonne Bay was where he fished for eels and the blue crabs that he loved. Once he bought me a B-B gun. As I fired the gun, a B-B bounced off a tree and struck him in the forehead. While I prayed to God, I was also looking for a place to be buried after he killed me. To my surprise he sat me down and began showing me how to handle a gun safely. I could not concentrate on the lesson, because I was asking myself, "Who is this man and where has my father gone?" I don't think Frank, Jr. was ever involved in an occasion when our father showed signs that he cared for the safety of his children. He rarely witnessed or felt our father's love and compassion. Even today, he has a deep hatred for our father and little appreciation for the name Frank that was proudly bestowed on our father's youngest son.

Chapter 6
Marital Adjustments and Death

I was now ten with a father who was still beating our mother. Suddenly, my parents separated, and he left our home. He moved into a second-floor apartment above the self-trained barber where Frank Jr. and I got our hair cut. On one occasion, after we received our haircuts, we went upstairs and paid our father a visit. Not long after we entered his home, he ran an errand and left us in his apartment. Being curious, we discovered a grapevine less than an arm's length from the windows. We picked the grapes and threw them at each other. Our father returned just as we were picking more grapes. He voiced his anger with words that I never heard in church and promised to punish me on my next visit because I was the oldest. As hard as it was for me to do, I finally went to his apartment and told him I was ready to receive my beating. Instead of punishing me, he laughed and complimented me on becoming a man. Soon after that incident he left Bayonne and I never personally saw or communicated with him again. Decades later, while I was in the Air Force, an unemotional Frank told me that our father had died and was buried in Baltimore, Maryland. However, New York's census records contradicted Frank's information. Those records indicated he died in 1963 and was buried in Brooklyn, New York.

Several months after the December 7, 1941, Japanese attack on Pearl Harbor, my mother and three of my sisters, Louise, Cornelia II, and Estella were stricken with Tuberculosis (TB). My mother received outpatient care while my sisters were hospitalized. Louise gave birth to Harold, the first of her seven children, while in the hospital and was then sent home. Connie gave birth to David and died shortly thereafter. Estella followed Connie in death. Stella and Louise had once competed for the affection of Sam Allen. It was Louise who lost the competition after marrying and later having to divorce the adulterous batterer.

Cornelia and Paul had shared very little time together. He was drafted into the army in 1942, the year she died. While her death was painful for the family, it was a devastating loss for Paul. The army had given him permission to bury Connie, but the thing I remember most was watching Paul on the corner of Eighteenth Street and Avenue C. He wore an olive drab Army uniform and was crying, shamelessly, in broad daylight. His display of grief caused me to have mixed emotions because my father repeatedly said, *"Real men don't cry!"* But I had seen my father cry when he got into a fight and hurt his finger. But I believe his pain may have been both emotional and physical because his opponent, *Blue*, was his best friend. At age twelve, I still wanted to calm and be comforted by Paul. Kids in our family did not attend funerals and while I wanted to see my beloved sisters one more time, I also wanted to retain my earlier visions of them. I had watched while they hosted 25¢ house parties. They were so full of life as they danced to lively music, hugged each other tightly, and laughed while having so much fun.

Chapter 7
Musical Influences and Preferences

Oscar's degree was in education with an emphasis in English. But the sun didn't always shine on what in those days was the highly respected field of teaching. A White child entering his classroom, politely said, "Good morning, Mr. N". Oscar believed that children are led by adult words and deeds to develop a sense of superiority while demeaning other races. Teaching was Oscar's profession, but music was his real love. Carnegie Hall, a musical center in New York City opened its doors in 1891. It set the standard for musical excellence and people of color rarely performed there. In 1944, Duke Ellington a renowned pianist, composer, and Black band leader, created 'The Carnegie Hall Concerts', a live album. The project opened the doors a little wider for colored performers. Oscar also performed there in 1944 with a well-received classical piano performance. Because of Oscar's influence, I enjoyed Mario Lanza's delivery of Nessun dorma even before the rise of Luciano Pavarotti, the unmatched and world-famous Italian tenor. Maria Callas' delivery of Mamma Morta and Madam Butterfly were superb. Was there anything that the operatic soprano Kathleen Battle couldn't sing? I also enjoyed Roy Hamilton, Tina Turner, John Waite, Whitney Houston, Luther Vandross, Beyonce, Barbra Streisand, Il Divo, Marvin Gaye, Aretha Franklin, Alicia Keys, Celine Dion, Teddy Pendergrass, Gladys Knight, Lionel Richie, Mariah Carey, Radiohead, Willie Nelson, Isaac Hayes, and others. As a matter of fact, I began building separate musical component high-fidelity systems in the 1950s. I wanted Tchaikovsky's 1812 overture to sound real with its speaker and amplifier destroying canons booming away. The overture honors the successful Russian defense against Napoleon in 1812. Piotr Ilyich wrote the overture in 1880, almost seventy years after the battle and it still inspires musical audiences today.

Frank's relationship with our father, was in sharp contrast to the one he shares with his son, Charles Lawrence Howard. I've watched, with a little envy, when Charlz, as he chooses to spell his name, turns the simple task of cutting Frank's hair into a special event. Charlz vacationed with me in Dallas when he was thirteen. One day I heard him urgently calling Uncle Robert! I opened the door and saw him running up the steps. Charging shamelessly behind him were several neighborhood girls. If I'd looked like *my big kid,* whom I regarded as second to none in any group, I would have stumbled and let the girls catch me.

On The Boardwalk in Atlantic City. New Jersey

Chapter 8
Learning the Hard Way

My father took me fishing where logs were placed vertically into the soil beneath the water to keep the soil from being washed away. Tommy Lee, Sonny, and I had been playing at that west Twentieth Street area. Now, late in the day, they swam while I sat on the bokay, as the bound logs were called. When they left the water, I foolishly chose that moment to jump into the deep water. But I had never learned how to swim. Tommy Lee, noting my panic, quickly dove back in and saved me. There were no lectures from my friends on the way home, but the next time we went swimming, it was after we had walked across the Bayonne bridge and paid ten cents to enter a public pool in Staten Island, New York. That is where they took turns teaching me how to swim.

Sonny, his brother, and I once went to a penny arcade site located on First Street. Between games, I went outside just in time to see Howard darting around Second Street. As I got close, I heard a man shouting, "What are you kids doing?" Sonny took off running. I didn't know what was going on, but I ran too. I wasn't worried about being caught because I prided myself on being the fastest kid on the block. In fact, the kid's had nicknamed me Streamline. Suddenly, a hand lifted me off my feet. Sonny's brother was also caught, so Sonny stopped running and returned to where we were being held. The men were about 20 years of age and I feared we would be taken to jail. But they took us to Sonny's house and told his father what had happened. That was a decision that kept me from having a police record and allowed a yet unknown security door, very important to my future, to remain open.

Chapter 9
The Extended Family

My sister Ruth met and married Leon Borroum, a seaman stationed at the Bayonne Naval Base. Years later the base provided employment for two of their daughters. Lois served as an executive secretary. After retiring from her civilian position, she got a job at Saint Peter's College in Jersey City. While there, she took college courses, graduated with honors, and retired again. Leslie worked as a civilian transportation assistant. Then, after serving a tour in the Army as a WAC, she continued to work for the army as a civilian in Norfolk, Virginia. She worked there as a cargo booking agent for many years before retiring. The sisters, now economically independent, once shared a home together in Bayonne.

The Borroum Sibs—Val, JoAnn, Lois, Leon Jr, Joe, Leslie, Ted, Ken

Liberty Park in Jersey City was a place for some of our large local family gatherings. It was there, that we dined on barbecue and picnic delicacies while viewing the Statue of Liberty and telling stories.

New Jersey and New York Ayers Descendants
With Their Significant Others

Nieces And Nephews In The Jersey City Hood

Nationwide reunions were much larger than local events and were carefully planned by elected committees within the state where the reunion was to be held. Over 200 Ayers' family members met annually in places like Huntsville, Alabama, Chicago, Illinois. Indianapolis, Indiana,

Atlanta, Georgia, Paducah, Kentucky, Philadelphia, Pennsylvania, Memphis, Tennessee, and elsewhere. Obie and Earnest Ayers were among our many fine family leaders. But when I wanted to know about past family activities, I asked Johnnie Mae Miles or Handsome Ayers. They were walking encyclopedias on family history.

On the Friday before Labor Day, the national reunion hosts prepared meals that featured fried fish. On Saturday there was a picnic with barbecue, salads, desserts, and children's activities. Saturday evenings were reserved for a business meeting, to choose the state and the host committee for the next nationwide reunion. That meeting was followed by evening snacks, music, and dancing. On Sunday, the family attended church as a group. An ordained Ayers minister usually took part in the services. The church services were followed by a semi-formal attired full course meal at a pre-designated site. That meal marked the conclusion of planned activities, and we began returning to our homes.

During the 1988 downturn in the nation's economy, my cousin Michael Porter came to Houston and stayed with my family. He was a mechanical engineer who had accepted a job at the relatively close nuclear plant in Bay City, Texas. Later, he and Lisa, his wife, moved to Bay City where the plant was located, before eventually returning to Alabama after the economy improved. Michael served as the president of the host committee for the 2017 Ayers family reunion in Louisville, Kentucky. During the pandemic years of 2020 and 2021, our national family members met in zoom or virtual meetings.

YoungAyers Nationwide Family Reunion Group

An Ayers Nationwide Family Reunion Grouping—
Photos Compiled By Cousin Johnnie Mae Miles

Chapter 10
Suppressed Memories

There was a body of water that ran the full length of Bayonne and Jersey City on their West sides. This was a time when President Franklin Delano Roosevelt delivered his famous speech regarding the Japanese attack on Pearl Harbor. He defined the attack as, A Day that will live in infamy and declared war against Japan, Germany, and Italy. My former abusers were all old enough to be drafted into military service. Everywhere you looked, there were posters of Uncle Sam, dressed in his red, white, and blue costume with his finger-pointing and the

15

words I want you on the poster. The posters were successful in sparking the flames of patriotism because many men volunteered for military service.

There was a beach about a quarter of a mile from Hudson Boulevard, on the Southwest side of Jersey City. We had to travel about a quarter of a mile from the Boulevard through some very tall brush to reach the water. The site which could not be seen from any paved areas came to be known as bare ass beach. I went to the beach alone one day and joined a group of kids who, like myself, were swimming in the nude. As we lay on the shore, joking with each other and enjoying the subtle breeze, a grown man invaded our space. He was a rather big and burly man with thick red bushy hair. Although he did not undress, he began to dominate the conversation with stories that were more and more sexual in nature. I became uneasy and decided to leave the gathering. As I travelled through the tall shrubbery, I heard a thrashing through the bushes sound. My something is wrong antenna rose, so I found a depression and hid as well as I could. As I lay in the depression, I saw the red headed man moving quickly in all directions. Finally, he headed back to the beach. I remained hidden until I was sure he was gone before I got up and reentered Bayonne. That incident lingered in my mind, and that night I had a nightmare.

I dreamed I was at the bottom of a huge, round, cone-like structure that became very narrow at the bottom. Spaced above me at intervals of about ten feet, were a series of circular ledges. John Bronowski, my hide and go seek abuser, stood above me dressed in a red, white, and blue costume—Was I afraid of Uncle Sam?—but he was bigger than his normal size. He repeatedly shouted in a booming voice while pointing his finger at me and saying, "I want you!" After I woke up, I realized that the trauma of my earlier encounter with pedophiles had affected me more deeply than I, as a child, could have imagined. How long would my insecure feelings last? But I had learned to sense danger and believed I could avoid being victimized again. The fact that I had shared those horrifying experiences along with others, contradicted the *misery loves company* remarks I'd often heard others say. Only time eases the lasting effects of having been abused. As a matter of fact, when my nephew Charlz attempted to embrace me, in a gesture of friendship, my instinctive avoidance of his embrace bothered him, and he told me how he felt. Other than my sister Eileen, he was the only family member back then that learned I had been sexually abused. One time he borrowed several thousand dollars from me. Many months later, I answered my doorbell only to find Charlz there. He had come to Texas, unannounced, from New York to repay his debt in full. He also tried to pay mafia type interest. I didn't accept the interest, because his demonstration of love for the *Kid*, as I used to call myself, was worth far more than the money. More important to me was the fact that a man could love another man without it being a sexual thing.

Before she died, my always protective mother told me to be careful of who I let stay in my home. She warned me that some relatives would disrespect my wife. Years later, a nephew bragged to me about having had sex with real virgins. He was unaware I knew he was talking about his pre-puberty aged cousins who were also my nieces. I told him raping children would never count as a consensual conquest and that I was not impressed by his criminal act. His verbal response of "allegations, allegations" were the last words I heard him say because I never spoke to him again. Abusers use threats of harming their victims or someone the child may love to cover up their shameful deeds. I know my revelations may cause some relatives to feel uncomfortable, but I will not be silent if I can protect others from abuse.

Chapter 11
Early Adolescence

I entered Junior High School with low grades. Instead of studying, I sold newspapers after school and then delivered bottles of milk later that night. A counselor decided I should become a clerk-typist but breaking my wrist while playing sandlot football kept me from learning how to type. A fire destroyed our home in 1945 and my brother Eddie moved us into the home he shared with his wife, Ethyl. Eddie, who was now a chemist, created and sold cosmetics from Paramount Studios, the name of his business.

I was completing the first semester of the ninth grade, when my algebra teacher sat me down for a one-on-one talk. She informed me that she would give me a passing grade if I did not take her class the second semester. I was gleeful rather than offended. I don't know how she pulled it off, but somehow, she managed to have me transferred to another algebra class. As I looked around my new room, I noticed two empty desks in the back of the room, where I could sleep. But on the other side of the room was a magnet-like attraction that caused the puppy dog love juices in me to rise. It was a girl with a beautiful brown complexion. Her face had been put together by an award-winning artist while her body seemed to glide when she walked without a hint of pretended sexuality. To me, she was a picture of virtue and the most dignified version of what every nice girl should be. I just had to find a way to get her to notice me. But getting Hazel Jamison's attention, or H. J. as I shall call her, was a task that was easier said than done. This was especially true since an attention-getting approach, I had tried in the recent past was a dismal failure. I was walking on a fence toward a porch, where a girl I liked was sitting. Before I reached the porch I fell and a dog that was in her yard bit me. I still have a scar on my upper lip where the dog's fang ripped it apart. Obviously, there were better ways of introducing myself. I finally decided to try and impress H. J. by showing her that I was a good student. Even today, I wonder how that idea entered my head? Was I as crazy as my brother Eddie had often stated? Her grades were A's, while my highest grade was a reluctantly awarded C. Was the hole I knocked in my head when I fell off that fence still there? How else could the thought that I could compete with her intellectually have penetrated my—I never studied—thick skull? Never-the-less, I was determined to win her.

For the first time, I began to study hard, and my grades improved noticeably. Even though my surprised teachers reported my progress to the whole class, H. J. did not notice me. But I was not ready to give up and I continued to study. My next report card was the best my family had ever seen; I even earned an A. Still, the only time I saw H. J's face light up was when the son of the barber who cut my hair, came around. All that studying was wasted, and my plan had failed. I resigned myself to accept the fact that love wasn't in the cards for me, and I might as well go back to sleeping in the back of the room. But my efforts to impress H J had gotten the attention of my teachers. They now knew I was capable of learning and pressured me to continue doing better. I also noticed that some classmates seemed to respect me more as my grades improved. I liked that. Apparently, my method of combining related understandings together improved my problem-solving skills. The young man who graduated from Bayonne High School, three and a half years later was a significantly better student than the one a desperate ninth-grade algebra teacher could no longer stand to teach. My later life would have been so different had I not been inspired to grow academically at that critical mind developing point in my life.

At seventeen, I was still delivering newspapers in the evenings. Normally, a truck would drop my papers off in my area. However, due to a change in procedure, the truck began delivering the papers to a central drop-off place where all the paperboys could pick up their loads. I had to travel a few blocks through an area reserved for a boy named Eugene. One day as I was carrying papers to my area, I was stopped by Eugene Grant and Nathan Blalock, one of his buddies. He accused me of selling papers in his territory. I tried to tell him I only delivered papers to the homes of my customers and did not sell to people on the street. But he was not interested in anything I had to say and called himself punishing me by hitting me in my left arm to make me drop all my papers. I left without fighting, but Eugene would face more than a toy bat when he and I later fought while we were alone. The street code was fight back or get beat up until you do. Suddenly, Eugene disappeared, and I learned he had joined the Army. Also, the three musketeers became one, because Tommy Lee and Sonny had also joined the Army in 1947.

I am glad I continued my friendships with Richie Pavlov and Sammy Jaworski who were White. We frequently hung out together. During our senior year Richie, Sammy, and I, along with Sammy's cousin Walter, played hooky. We took the subway to New York and went to the Paramount Theatre. One performance featured the movie, Indian Love Song, starring Nelson Eddy and Jeanette McDonald. We also saw a live performance by the Nat King Cole Trio. We enjoyed both shows but as we were leaving the Paramount, one of the guys sounded an alarm. It appeared we were being followed by a gang of rough look-ing kids. Fortunately, we also saw some police officers across the street. This was not the time to worry about getting a ticket for jay walking. We would have welcomed the jay walking attention. So, we hastily darted through traffic across the street and began following the officers. There appeared to be some justifications for our fears because the gang turned and ran in a different direction.

I graduated from Bayonne High School in 1948 and, once again, lost my closest friends. Their families could afford to send them to college, and they had a mindset whereby graduating from college was just another step in their progression. I, on the other hand, still on welfare, never gave college a second thought. I lost track of Richie, but I learned that Walter achieved success as a business executive while Sam became a commissioned officer in the Army. Walter and I talked to each other at our 1998 fiftieth high school class reunion. A six feet two-inch Willie Tomasevic, who I hoped didn't remember me from the first grade, was also in attendance. H. J. had been happily married to her school boyfriend for more than forty years and I was happy for them. She, her cousin Christine, and Ruth, my sister, were good friends. After earlier attending Friendship Baptist Church, we talked about the past and I thanked her for being the spark that inspired me to show my teachers I could do much better. Her young adult daughter and I were both shocked when the still pure H. J. innocently kissed me on my cheek.

Chapter 12
Post High School Challenges

I got a sample of what life for unskilled workers would be like. In my first job I tore down old buildings. I had only worked there three days when Buddy Hurst, a star football player, told me he had never seen anyone look so tired. My next job was in a laundry where I put dirty, and nasty smelling clothes into huge barrel shaped bins of hot waters. The job was more torturous on hot summer days. My third job was in a Jersey City thumb tack factory. I was given a twelve-by-eighteen-inch rectangular tray that had evenly spaced tack-head sized indentations in it. I shook unpainted tacks into the indentations until the points of the tacks faced up. Imagine doing that for twenty years. A co-worker then pressed the points of the tacks into small oblong shaped pieces of cardboard and painted them. He quit shortly after I was hired so my boss, Melvin, assigned all his brain challenging duties to me. After Melvin discovered my clerical background, he also had me keep records on the amount of work that was done by each of my coworkers, as it was the basis for their piece work salaries. He even had me, at eighteen years of age, open the factory in the morning. These significant increases in my job responsibility, were not accompanied by any increase in pay. Despite these dead-end jobs, I still did not visualize or consider a better future for myself. My periodic contacts with Tommy Lee and Sonny indicated they were not happy in the Army. While I didn't want to be drafted into the Army, former health concerns and a free physical, rather than patriotism, motivated me to voluntarily join the Air Force. Suddenly, unlike my brothers, I passed my physical examination. On November 23, 1948, I reported to Lackland Air Force Base in San Antonio, Texas to begin my basic military training. President Truman had recently signed executive orders 9880 and 9881. The orders would require time to implement and were intended to reverse President Wilson's earlier directives that had segregated federal workers and the military services.

A two-story building with very large open bay areas on each floor was my home for the next ten weeks. The barracks had a private room reserved for our flight marcher. Each large open bay area held bunk bed type cots. Those double stacked beds were placed side-by-side several feet apart on both sides of each floor. At the end of each bay, there were tiled areas containing side-by-side toilets and showers without partitions. All the recruits in my flight were Black and appeared to be between eighteen and twenty years of age. But one recruit stood out because he was twenty-six.

During a 1940 decade when the height of a five-foot nine man was thought to be fine, Corporal Taylor, our flight marcher, was a six feet two-inch, compact male, without an ounce of fat on him. His dress was perfect and when you heard his voice, there was never any question about who was in charge. When I entered the Air Force, enlisted men were ranked by grade. Those grades were private (Pvt. Enlisted man level one or E-1), private first class (Pfc. E-2), corporal (Cpl. E-3), sergeant (Sgt. E-4), staff sergeant (S/Sgt. E-5), technical sergeant (T/Sgt. E-6) and master sergeant (M/Sgt. E-7). Enlisted men from Cpl. through M/Sgt were also classified as NCOs or non-commissioned officers. Warrant officers are NCOs appointed by warrants. They generally are specialists in technical fields. While warrant officers outrank other NCOs, they are below commissioned officers in the rank structure.

I was a little surprised at how quickly the men formed friendships with each other. Our twenty-six-year-old airman, however, felt his advanced age qualified him to be our leader. He was extremely dark, and we referred to him, among ourselves, as Flash-black. Soon, he began bullying airmen who would not listen to him. I caught his attention as I was sitting on my wooden footlocker that measured 32" long, 16 ½" wide, and 13 ½" deep with eight sharp corners. I was writing to tell my family how much I missed them. Flash-black, picked this time to ask me to box him. I tried to decline, but he wouldn't go away. Suddenly, I felt as if I was being faced by both Eugene, my newspaper competitor, and Johnny B, my hide and go seek child abuser. Now they were before me, seemingly combined into one person. When I told Flash-black, we would not be playing, I was really directing my long-suppressed anger toward Johnny B and Eugene G. Never-the-less, men don't issue challenges without hearing from eager witnesses. "Fight, fight", was the battle cry from the spectators. Without hesitation, Flash-black began his attack. He was a better fighter than I had imagined, and I was no match for his boxing skills. However, I had delivered milk and newspapers to customers for more than five years and that strength and endurance allowed me to stay close and neutralize his boxing skills. Finally, I tackled him and knocked him to the floor. I arose and raised my footlocker over my head. I turned the locker so that I could stab him with one of its corners. The footlocker missed his head, but it struck his body and stunned him. Jumping high into the air, I landed on his chest with both of my knees. It was a move I had seen in the movies. I then grabbed his foot and started dragging him over to an upstairs open window, fully intending to throw him out. It was at this point that several trainees grabbed my arms and pulled me away. After Flash Black regained his footing, he did not leave the scene, but he also did not attempt to resume the conflict. As he and I finally turned and started to move in different directions, I heard someone say, "Nobody likes a bully Jersey. You did good". But it troubled me to realize how close I had come to trying to kill a human being. That wasn't the kind of person I wanted to be.

Cpl. Taylor wore a blue and white plastic helmet and when a trainee messed up, he would hit them over the head. Shortly before our training flight was to be given permission to go into town, I did something that displeased him. One of the guys told a joke that caused me to laugh out loud. Cpl. Taylor did not hide his irritation and the top of my head felt the force of his helmet as he landed a painful blow that hurt both my head and my pride. Days later I went to town and did an imitation of my father by pursuing girls and drinking too much. I then went to the United Services Organization (USO), a recreational area for military personnel. As I looked around the room, I saw Cpl. Taylor. He was seated beside a young woman. The loss of self-esteem since being hit over the head, coupled with an over consumption of alcohol caused my brain to stop working. Suicide was never a conscious thought, but there must have been a death wish lurking nearby. What else could have caused me to sit down at his table without being invited? This was a time when I needed a miracle; I mean the one where two orderlies in white jackets dragged me back to an insane asylum. Unfortunately, this was reality. and I shouted about how ugly his date was and how little rain-drops would make her short hair appear to be even more messy. Then I invited him to go outside. No one was surprised when he accepted my challenge. I don't remember what happened next, but my friends later told me that as I began to raise my fists, he hit me so hard that when I fell to the ground, part of my head slid under the bumper of a car. When I regained consciousness, after my alcoholic slumber, I was in my bed. Cpl. Taylor had driven me back to the base. As if all that had happened wasn't bad enough, I realized I could be punished for insubordination or failure to respect a person of higher rank. So, on the same day that I was physically with a woman for the first time, I got knocked out cold, as my father would say, by a real man.

As I was thinking about my problems a trainee entered the bar-racks and insisted that everyone listen to him so he could tell us about the Intelligence test we'd all taken. He had bragged that he planned to apply for officer's candidate school or OCS. Applicants for OCS were required to have received an intelligence quotient score of one hundred and ten or better. He could hardly contain himself as he yelled, "I made it. I scored over 110 and Jersey, (referring to me) you scored higher than I did." I don't know how he got access to the information, but an average IQ score has been defined by some as a score between 85 and 115. According to statistics, 68% of the people tested score in that range.

Flash Black's next encounter was with a trainee we called Tex. Our leader assembled our training flight outside the barracks. After sending Flash-black and Tex back into the barracks alone, he marched all of us away. When we returned, Tex and Flash-black were washing away signs of their struggle. Neither of them reported what had happened, but Flash-black was now less inclined to bully others. The more I thought about my hotheaded actions, especially when viewed in the context of Taylor's handling of the fight between Tex and Flash-black, the more I realized that for Taylor it was just another day at the office. He had forcefully placed into my understanding, with one devastating blow, the fact that you must bring more to an encounter than unbridled emotions. Knowing your enemy's strength and responding accordingly, is what helps soldiers survive in times of war. Basic training was finally over and my flight marcher, did not charge me with a violation. I too was promoted to Pfc. And upon further reflection, it appeared that the objectives of basic training were so much more than just a series of repetitive drills, exercises, and encounters. The training was a lockstep approach to self-discipline that took individuals from all walks of life, with differing points of view, and molded them into unified fighting forces that would follow orders without questions. What is amazing is that these seemingly tried-and-true transformations, that occurred over a noticeably short period of time, appear to work. Trainees were shaped into combatants who trusted each other and routinely engaged in united actions as if their lives depended upon them taking care of each other, because, in fact, it did.

Chapter 13
Black Restrictions

I felt the stares of airmen as I entered Topeka AFB. They were staring because there were no Black airmen on that base. I had mistakenly been sent to an all-White base to wait until the next intelligence class at Lowry AFB in Denver, Colorado began. Despite President Truman's desegregation orders, I could not be housed with White airmen yet. I had no idea when or even if I would be allowed to attend the class at all. The temporary solution to my unfolding dilemma was to have me, an enlisted man, occupy a room alone in the Bachelor Officers Quarters (BOQ). The next day as I was being transferred to a based in Salina, Kansas. the personnel director, who was a general, approached me. I guess he wanted to find out first-hand what a Black airman was doing on his still all-White base. Our brief conversation enabled the general to learn that I was from New Jersey. As he walked away, he told the sergeant to send me to McGuire Air Force Base in New Jersey. It was a base that had Black airmen and was only seventy miles from my hometown. That timely intervention significantly affected my ultimate destiny.

In 1949, McGuire AFB shared a common border with Fort Dix, an Army post near Wrightstown, New Jersey. Trenton, the capital of New Jersey, was nearby. Although I remained a candidate for intelligence school, I was assigned to the Black base service squadron. While the twice replaced commanders, leading the squadron, were White officers, the first sergeant who led the squadron was Black. Those men performed every menial task one could imagine. I say those men, because two Black airmen performed tasks outside the squadron. One was a Sgt who worked at Wing Personnel. The second airman was me. I was sent to receive on-the-job training (OJT) at Wing Intelligence.

My Officer-in-Charge (OIC) was Captain Mack. The non-commissioned-officer-in charge (NCOIC) was Master Sergeant Rothhammer. I spent almost the whole first day trying to type a half page TWX or electronic letter. I don't think my early work was valued very much because Captain Mack gave me three day passes to go home each week after the Thursday workdays. Some may question the value of on-the-job training, but Master Sergeant Rothhammer taught me a lot. In its most basic sense operational intelligence involved collecting, collating, and disseminating information needed by our men to successfully plan and execute their missions. He taught me how to research and fact check the data for accuracy. I learned to display data into formats where it could be used by our crews while airborne. It wasn't long before Captain Mack used more of my data in his briefings. The downside of my improved performance was that he stopped giving me three-day passes to go home each Thursday.

Our base facilities included an exchange store where clothing, musical items, toys, and games could be purchased. The food store was called a commissary. While the base library and movie theater were well attended, it was the military clubs that proved to be the favorite evening destinations for most of us. You had to be a commissioned officer, warrant officer, or officer accompanied guest to gain admittance to the officer's club. Because of Tuskegee pilot complaints, Black officers were permitted into the officer's club. The NCO club in 1949 was for enlisted men with a rank of Cpl. through M/Sgt. However, the Black airmen's club, for lower ranking Black enlisted men, was the club that was most favored. It had two adjoining parts that were formed into one club. One section had booths, a juke box and seating at the bar. The other section, joined by an enclosed walkway, had a bandstand, tables, and a dance floor. Many NCOs frequented the airmen's club because it was like a magnet for women.

Chapter 14
Offensive Humor Is Dangerous

Burlington New Jersey was where airmen threw caution to the wind to have a good time—my father would have loved the place. Once, a group of paratroopers from Fort Dix entered that café, which we in the Air Force regarded as our territory. Prior friction between the groups had never become violent before Airman McCullough was assigned to our base. He was what some might describe as a real pretty boy, but he would fight anyone who questioned his manhood. An Army paratrooper jokingly stated that McCullough was pretty enough to be a girl, and before he got the expected laugh, he was attacked by McCullough. The fight spread quickly as others got involved. A bottle grazed my eyebrow, and I was forced to defend myself. The fight didn't last long and there were no noticeable injuries.

Weeks later, I drove one of three '*deuce and a half (2 ½) ton military trucks*' loaded with airmen to Pennsylvania. Our squadron had been invited to a YWCA dance in the Germantown section of Philadelphia. I liked listening to the sound that a double clutched truck makes during gear changes. As we arrived at the YWCA, we saw a crowd of men around the building, but things remained peaceful. Later, Airman McCullough went outside for a smoke and started a fight. That caused other civilians to attack any airmen outside the YWCA. Airmen in the building rushed out and soon we were engaged in a massive street brawl. After the fight, we boarded our trucks and attempted to leave. One of the trucks caught on fire, but we managed to put the fire out ourselves and headed back to our base. As we were crossing the bridge between Philadelphia and New Jersey, a jeep filled with military police stopped our three-truck convoy. Some of the guys were ready to take on the military police, but we avoided further conflict. A group of unhappy officials greeted us at the airbase when we arrived, and we were all restricted to the base indefinitely. That restriction ended sooner than expected because all base units and equipment were soon moved to Barksdale AFB in Louisiana. The previous occupants of the base were moved to another base. Black airmen were concerned because of graffiti on walls saying what the White airmen planned to do when they got those *N's* down south. All the units remained intact initially, and as Blacks sat in their segregated dayroom, the telephone rang. The caller asked to speak to some airman. When told that person had probably transferred, her next question was, "What's your name?" Instantly a long line was formed to the rear of that phone. We also noticed a message near the phone advising us not to raise the price of dating because the ladies did not charge anything.

Chapter 15
Brutality And Good Times

It was a nice day in Shreveport, when I without a care in the world, was enjoying the weekend. The temperature was mild, and the breeze in the trees seemed to be in tune with the birds tweeting melodies. The Black airmen walking before me also appeared to be enjoying the day. Suddenly, two White policemen got out of their car and approached one of those men. One officer demanded that the man singled out, do the *hambone*. His hip reply of, "Man I don't know no hambone", angered them and one officer immediately began swinging his night stick indiscriminately at the head, shoulders, and other areas of the young man's body while shouting, "Boy don't you sass me!" The incident was over almost as quickly as it had begun. The officers, who had momentarily found a diversion from their boredom, returned to their car while the young man struggled to rise. Few in the barracks were shocked by my report. The Juba dance or hambone was originally known as Patting' Juba. It is a style of dance involving stomping as well as slapping and patting the arms, legs, chest, and cheeks in a musical pattern or cadence. Other acts of brutality were discussed that day, especially driving while Black. The DWB incidents involved Blacks, who were justifiably offended after being stopped and punished simply because they did not act subservient enough when responding to demeaning police questions.

However, Louisiana was also a place where we often heard the expression, *Laisses les bon temp rouler* or the French way of saying let the good times roll. One evening after I left the Chicken Shack, a nightclub, I hurriedly turned a corner and bumped into a young woman. Once again, I was mesmerized, just as I had been when I saw H J, in my algebra class. This time, I was determined to win her over.

I started out by apologizing and then asked if she could tell me where the Chicken Shack, the place I had just left, was located. It was corny but, as I had hoped, she indicated she was heading in that direction and would guide me part of the way. Before we got to the club, she stopped at a house and sat on the porch. After hearing my repeated pleas, she relented and gave me a telephone number. All I could think about afterward was Tommie. The next day, I hurriedly called this Creole dream girl. The voice answering the phone was not Tommie's, and after a few exchanges she was about to hang up when I heard someone laughing in the background. That laughter was followed by Tommie's voice. She had given me her friend's phone number and I discovered that neither she nor her friend lived in the house where she had taken a seat on the porch. How should I describe Tommie? Her face had been sculped by an artist possessing Michael Angelo's talents. Her perfectly aligned teeth glistened within the framework of her 'oh so' kissable lips. I'd met and enjoyed myself with women who didn't fit my old, brainwashed description of beauty, but Tommie was all of that and more. She was five feet six inches tall and weighed about 125 pounds. Her light complexion, long silky hair, small waist, sexy hips, thighs, and legs completed heavens carefully designed masterpiece and *I was in love…again!*

The Black airmen were soon integrated with other races, and I slept on a separate cot between two White airmen. Wash basins, commodes and showers without partitions were the order of the day and, at least for me, conversations and even joking remarks flowed naturally between the races. There were also areas on the base where people could hunt safely, and that privilege had been extended to the sheriff of Bossier City by the previous base commander. The new base commander withdrew hunting privileges on base by non-military personnel. A story began to circulate that Bossier City police cars lined up outside the main gate where they could harass military traffic. But I had fewer reasons to visit Shreveport because I found a letter on the floor by Tommie's bedroom from a husband I didn't know existed. The decision to leave her was made for me, because I was now scheduled after integration to take formal operational intelligence classes at Lowry AFB in Denver, Colorado.

Like the rest of the individuals, who would be my classmates, I was entering into a racially mixed environment. What I did not expect to find was a group of English-speaking soldiers from other countries such as Turkey. Our foreign allies had been enrolled in American classes during the time when I, an American, was denied admission because I was Black. One exchange between a Turkish soldier and me, lingers in my mind. It was about the combat expectations that Turkey levied upon its soldiers. He stated each soldier was expected to kill ten enemy soldiers. When I asked what would happen when he reached his quota, he responded, rather casually, that he helped his friends. I'd heard stories about the ferocity that our Turkish allies displayed in Korea, and I was inclined to believe him.

Initially, my classmates and I listened to music and played games in the barracks. Chess fascinated me, and I learned to play the game by watching. I won my first game, and from that point on, I was hooked. I also noticed that my love for Tommie was fading. In a way that was a good thing because I needed to pay attention to my challenging classes. Most of my relationships with my White classmates seemed fine until they learned at the end of the course that I had placed second in the class. Some of them blatantly revealed their inner beliefs that Blacks lacked the intelligence to mentally outperform Whites. Some tried to get an investigation of the matter started. In settings where opportunities are afforded to all students equally, classroom standing will be more a measure of one's work ethic and reasoning processes, rather than the color of their skin. I think my advantage was M/Sgt Rothhammer's intense on the job training. The protest was soon dismissed. In matters of the heart, Tommie had left Shreveport to be with her military husband and I received orders transferring me to Japan.

Chapter 16
Do Lieutenants Have That Much Authority?

Hamilton AFB in California was the port of departure for airmen going overseas. Some of us left the base and went swimming in a bayou on private property. Suddenly trucks accompanied by military police appeared. Swimming there had been going on for some time and a lieutenant decided to discourage future violations. I don't know how the others were punished but my assignment to Japan was cancelled and I was sent to a U. S. protectorate in the east Pacific called *Guam*. In addition to Anderson AFB where I was stationed, Guam had a Naval base and an Army complex. It also had some Japanese soldiers who still refused to surrender even though WWII had ended five years earlier in 1945. Guam's extreme weather patterns, varied from Sahara Desert like heat to Niagara Falls like torrential rainstorms.

Airmen lived in large open bay Quonset huts. A Quonset hut is a lightweight prefabricated structure made of corrugated and galvanized steel that has a semi cylinder-shaped cross section. Our bathing areas and toilets were in a separate prefabricated wooden structure equipped with partition-less wash basins, commodes, and shower facilities. If you were shy and bashful, your choices were to man up and accept the norm or use the nearby bushes that covered the ground everywhere. Malaria carrying mosquitoes were a big problem, so each sleeping cot was fitted with a protective net. My first sergeant, M/Sgt. Johnson, was a hulking, wrestler sized man. He was also housed in a Quonset hut before his family arrived. Robby and Danny were Black airmen who lived in my Quonset hut. I had been promoted to Cpl. at Barksdale AFB and except for Corporal Barringer, a native of North Carolina, there was very little openly displayed prejudice. Cpl. Harris, with whom I was to work in intelligence, warned me not to get into an encounter with Barringer. You might say Barringer was the Caucasian version of Flash-black. Yes, he called me the N word. No, I did not fight him. Yes, I watched as he got his butt kicked by Cpl. Sullivan, an Irishman, who didn't like people speculating about what his wife in America, might be doing. Sully's battering message really spoiled the day for a person who seemed to hate everybody. The lesson for me was if you're not being physically attacked, ignore racially motivated insults.

In 1950, it was hard to meet Guamanian women. The strong Catholic missionary influence insured that the girls were almost always chaperoned. The only thing I had to look forward to on a regular basis, was Major Hamilton's daily butt chewing. I learned to make some harmless mistake early in the day. After that, I could blow the place up and he wouldn't bat an eye. That changed after July 1950 when our military forces were actively engaged in the Korean War. General MacArthur led relatively inexperienced American forces deep into North Korean territory. Those forces were nearly pushed into the sea before our massive reinforcements enabled them to establish a beachhead around the port of Pusan (now named Busan). I performed temporary duty (TDY), in support of the Korean war from a base in Northern Japan. Our mission was to provide air rescue operational support to our Korean forces. Initially, the prospect of going to Japan and dating girls was something **we** eagerly looked forward to doing. However, our focus was redirected to the successful completion of our life saving air rescue missions. I'm not sure how some would define me, but if doing all I could to save our men was patriotic, then I earned the title. The Korean War ended when our President, Dwight D. Eisenhower, a former Five Star General, pressured North, and South Korea to sign a peace treaty on July 27, 1953. The war had started due to too many reunification disagreements. It was a war which lacked any massive military battles, still it killed almost forty thousand Americans and more than a million Korean and Chinese fighters. Even though I never set foot on Korean soil, supporting forces stationed in Japan were decorated with the Korean Service Medal. However, Albert Mason physically served in Korea. His Army unit was overrun in November of 1950 by Chinese forces near Unsan, North Korea. The attacking Chinese were blowing bugles and hurling grenades as they advanced. Albert was left for dead but, amazingly, was his unit's lone hospitalized survivor. My sister Eileen had enlisted in the Women Army Corp (WAC) of the United States Army and worked as a nurse at his medical treatment center. They were attracted to each other and eventually got married.

Chapter 17
Gambling Fever

Near the end of my tour on Guam, I was scheduled for a five-day duty-free vacation in Japan. But a thief robbed me and most of the men in our hut as we slept. All I had left was the change from a purchase I had made earlier in the day with a twenty-dollar bill. Since I no longer had enough money to even consider going to Japan, I stayed in the barracks and gambled. The game was draw poker with deuces wild. To this day, I still can't believe how much I won. It was enough to finance my trip in lavish style. I also lent some money to a friend. Upon arriving in Japan, we checked into our housing at Johnson Air Base. But this was not a day for us to go to the mess hall where the one meal fits all diners were spooned onto a metal food tray. I wanted to excite my taste buds with the scrumptious flavors of superbly prepared Japanese meals. Even more exciting was the prospects of dating girls. I really enjoyed being in Japan and my joy would continue because I was scheduled to end my time in Guam. I had also been promoted to sergeant and because sergeants didn't have to do kitchen chores, I persuaded the first sergeant to let me re-type the duty roster and remove my name. Stella was probably smiling in Heaven.

My next duty station was Westover AFB near Springfield, Massachusetts. It was 136 miles from my home. Before Eddie died of a heart attack in 1951, both he and Oscar had moved our mother into a single-family dwelling on Fifth and Monmouth in Jersey City. I was driving to that home after devoting my best efforts to support the Korean war. Now, as a returning war veteran, I felt patriotic and proud. Just as I was turning slightly to the right off Newark Avenue, my mother's house was coming into view. It was then that I saw the flashing red lights of a police car behind me, and I stopped. Before any words were exchanged, I saw an officer, who was approaching, holding his unholstered pistol through my side-view mirror. When he spoke, he told me I fit the description of a robbery suspect. After I showed the policeman my military orders and pointed to my mother's home, he holstered his gun. Even so, his reason for stopping me was an unlikely excuse. It was dark. How could he identify my features? The so-called robbery he referred to had been committed over 20 miles away in Newark. I guess he was just acting on the belief that all Black men look dangerously alike, especially if they are driving in a predominantly White residential neighborhood. That event let me know that even ardent supporters of the law are not safe in America. The deeply imbedded suspicions about Negroes, when coupled with assumptions of guilt, and the fear of resistance puts Black men in unwarranted danger even if they very politely ask an officer why they are being stopped. I now realized how lucky I had been to have cluelessly survived encounters that I had unwisely believed occurred only in the South,

Chapter 18
The Wild West

The work at my 5th Air Rescue duty site on Westover AFB was neither dangerous nor demanding. However, gambling was still an attraction for me. My system of wagering small amounts, then doubling up when I lost and never asking for another card that might cause me to lose had worked for me. In fact, I won with such regularity that I came to regard the money as a part of my monthly pay. My system no longer worked when the club limited the money that could be bet in a single hand. Since I could no longer double up and catch up, unlike my father, I quit gambling.

I received temporary orders to report to Mountain Home AFB in Idaho. The base was located near the sparsely populated city of Mountain Home. To enjoy any kind of night life, Black airmen travelled fifty miles to Boise, the capital of Idaho. This was an area where the residents considered it amusing to post signs saying *The Wind Will Blow Up This Road*. There was nothing funny about the signs when the fierce winter winds did blow and there was nothing to blunt their icy blasts. I volunteered my help at a horseback riding stable in town. I had previously worked around a riding stable in Jersey City and could take care of horses. The owners of this riding stable allowed me to participate in rounding up wild mustangs that were captured from across the Owyhee River. The river is a branch of the 280-mile Snake River in northern Nevada, southwestern Idaho, and southeastern Oregon. Capturing wild horses was a rare treat for me. I bought one of the captured mustangs for $25 and named her Maude. Since Elmer fed and housed Maude at no cost to me, it was sort of a form of payment for my help.

Maude was a bay-colored brown horse with a black mane and tail. I wanted to break her gently and started out by placing a car tire around her neck. I then placed an empty saddle upon her back and got her used to the bit in her mouth. Next, I let her feel my weight in the stirrups and then my full weight in the saddle. Jimmy, a White airman, was breaking one of Elmer's pink-skinned white horses. After riding the horses in a closed coral for several days without any effort on the part of either horse to buck us off, we left the ranch to find out how the horses would respond in unfamiliar areas. We had ridden about a mile when Jimmy imitated some rodeo riders by easing his body behind the saddle and leaning back until he laid on the rump of his horse. Now I knew better, but I could not stop myself. The event is recorded in my non-existent diary under the caption of do something very stupid today. As I was about to fully recline on Maude's rump, I felt her twitch. I quickly tried to pull myself back into the saddle, but by this time Maude was into a bucking routine that would have delighted any rodeo crowd. The third time I reached for the saddle horn, Maude veered left, and I found myself with nothing under me except air. My impact with the ground was sudden and painful. Seeing Jimmy laughing made the situation even worse. After running a short distant, Maude settled down and was grazing on nearby patches of grass. Fortunately, I was able to remount her and return to the stables. A week after that smashing return to reality, things really improved. The citizens of Mountain Home witnessed an unusual sight. It was me, decked out in my finest cowboy apparel and riding Rally, Elmer's prized black and white pinto stallion. The horse stood 16 hands high which is five feet and four inches from his hooves to his withers or shoulders. I'd never ridden a taller horse or one whose style had him lifting one hoof and then the other in such a rhythmic and high stepping fashion. I proudly guided Rally with his bobbing head and princely pace as we paraded pass the excited crowds.

The air force sent me to Mountain Home to do a job requiring the total restructuring of an operational intelligence section. The work had to be completed during the creation of a newly formed squadron. The classified material on hand was outdated and accountability methods were poor. The cataloging and destruction of items on hand was time consuming and frustrating. My subordinates, except for Cpl. Moore, had very little intelligence experience. Yet, we made progress. In the end, outdated materials were destroyed and replaced. And while Cpl. Moore's contributions were outstanding, he was discovered in a hotel room with another man. He was investigated by the Office of Special Investigation and declared unfit for military service. By that time, we had finished our work and I returned to Westover AFB. However, within months, I was on my way to another duty station.

Chapter 19
Career Indecision

My four-year enlistment was due to end in 1952. But the Korean War had not ended, and I was facing an involuntary extension of my four-year enlistment. The air force offered financial bonuses to men who would reenlist. I decided that since I wouldn't be discharged on my discharge due date, I might as well reenlist and be paid the bonus. Months later, the air force discontinued its policy of retaining men involuntarily. By then, I was on my way to Yakota Air Base in Japan. The squadron my office served flew RF-14 reconnaissance fighter jets. Our intelligence officer had been a minister prior to joining the Air Force. He left control of my subordinates to me and used the information we produced along with our photo intelligence section's data to brief our commander and his staff.

When I returned to the United States I was once again based at Westover AFB in Massachusetts. This time I was assigned to Group level Intelligence. At group level we inspected units that were subordinate to us in the chain of command. It was also at a time when Intelligence Specialists and Technicians were ineligible for promotion throughout the air force. My officer-in-charge, Major Seay, praised my work and urged me to apply for OCS. Unlike the IQ test I had taken in basic training, the OCS test contained significantly more science and math. I hadn't taken math since I was in the ninth grade, and I failed the OCS test. I later found myself in an awkward situation in which I, a group level S/Sgt (E-5) inspected the work of a squadron level M/Sgt. (E-7). I discovered some discrepancies in his classified material accountability procedures. They were issues where corrective action could have been completed in hours. My later follow up inspection showed the erroneous procedures were still being used and I noted the discrepancies in writing. The ride back to my temporary quarters on Selfridge AFB being driven by the M/Sgt I had just written up, was awkward.

Two days after my return to Westover AFB, my clerk typist had violated an article of the uniformed code of military justice (UCMJ). He was given an article fifteen hearing by our group commander and was reduced in rank from S/Sgt to Sgt. That same weekend the group scheduled a special training exercise. Usually, I stayed off base even though I had a private room assigned to me in the barracks. When the charge of quarters came through the barracks to awaken the men, he overlooked my room, since I was rarely ever there. I overslept and failed to get to work on time. I too was given an article fifteen hearing. While I remained a staff sergeant, because my OIC spoke highly of me, the colonel said I would never be promoted while I was in the group. Promotions in my field soon reopened, but true to his word, I did not get promoted. Once again, I was near the start of another reenlistment period and I put in for a transfer to another base, which was an option in those days. My OIC and the personnel officer tried to get me to stay. They even told me that our group commander was being involuntarily released from active duty because he had falsified some information on his application for reinstatement to active duty. But my reassignment was approved, and I was transferred to Florida.

Chapter 20
Death and Relocation

I was promoted to T/Sgt. Shortly after being transferred to MacDill AFB. My supervisor was a M/Sgt. Most of the time, we signed classified documents, produced by others, in and out. One friend that I met on base was Billy Martin, who worked as a supply clerk. I also befriended a married airman who lived off base with his wife and child. Frank was the type of person I usually avoided. For starters, he used drugs and was needlessly loud. Never-the-less I found myself being drawn to him. He was, almost the spitting image of Tommy Lee in size, and appearance, and, after you peeled away a layer of his bluster, he was quite intelligent. That, however, is where the similarities ended. Whereas Tommy Lee had been a thoughtful and prudent advisor, Frank was brash and a bold liar. I rejected his invitation to join him in the use of drugs, but it was Frank's behavior as we gambled that caused me to think of Lloyd Price. Lloyd sang a song entitled Stagger Lee. The song was about a dice shooting gambler who upon shooting seven would swear that he threw eight. I never saw Frank date outside of his marriage, but he knew where the girls hung out. My first date was with a woman who rightly deserved the name Black beauty. She was extremely dark and had won a local beauty contest. That relationship ended the second pay day after my arrival in Florida. She didn't believe my pay records were still in Massachusetts and that I hadn't been paid. Obviously, she wasn't impressed by my lovemaking skills, **or** she wouldn't have threatened me over money. I didn't want to figh**t,** so I left and never returned. Frank just laughed.

Billy and I went row boating on Tampa Bay. Neither he, a native of Manhattan, nor I were experienced with boats. One minute we were relaxing and enjoying our boat as it drifted slowly in the bay. It was so peaceful and relaxing until suddenly, the boat's movement increased, as if propelled by windblown sails. We tried to return to the area where we entered the water, but we were still being pulled away from the shore and into the Gulf of Mexico. Despite our panic, we decided to head toward a place on the shore whereby we'd be rowing across the current rather than against it. This effort paid off and we reached a distant but safe place on shore. Although we were tired, we decided to have a drink before returning to the base. As we were finishing our drinks, two young ladies entered the club. The ladies were stunning, and one of them was a close friend of Billy's future wife, Micheline.

The person who had dragged himself from the Tampa Bay shore, was both mesmerized and fully rejuvenated by the sight of the other female. She was all woman in every sense of the word. Being very dark, she was a departure from my brainwashed preference, but it was evident why Elaine had also won a beauty contest. Those feelings of being tired vanished. She was stacked! I silently asked myself what is going on? Before enlisting in the Air Force, I couldn't buy a date in my own hometown. Now, in Florida, I was chatting with my second beauty pageant winner. I discovered there was much to learn about this woman. She was a nurse who also sold insurance. She did not want or need my money. She was also a marvelous dancer, and the cha-cha moves she taught me gave onlookers the impression that I knew what I was doing. As our relationship got better, I thought we might be destined to walk down the aisle. But Elaine, initially unknown to me, had aborted our child and the deal-breaker for me was the fact that she never wanted to have kids. I'd lost Tommie because she was already married and as Elaine boarded the bus to visit her home in Tallahassee, I knew we were finished. Even though it was payday, I was in no mood to go gambling with Frank as usual. I was also disheartened by Elaine's' departure so I decided to return to my room on MacDill AFB. Sometime after midnight I found myself being awakened by the squadron's charge of quarters. He told me that our commander was on the phone and wanted to speak to me. I answered the phone, and my commander directed me to come to the Tampa Police Department because Frank had been killed. He did not give me any other information over the phone.

When I got to the police station, Ethyl came and put her arms around my neck. My mind was racing, and I assumed that Frank's tendency to be a bold gambler had finally caught up with him. As I was imagining what might have happened, a police officer came and escorted Ethyl away. The colonel approached me, and I was shocked to hear him say, "Don't get with that woman." He then revealed that Ethyl had killed Frank. I closed my eyes and muttered to myself that this could not be real. If ever two people loved each other, they would be named Frank and Ethyl. Frank had gone to town and gambled. Upon returning home, he appeared to be unusually disgruntled about an incident. He then directed his anger toward Ethyl. The argument that followed intensified and Frank began beating Ethyl. Ethyl, in desperation, grabbed a kitchen knife and stabbed Frank in his chest. After being stabbed, Frank made it out of the apartment and was leaning on the concrete post next to the stairways. He was still alive according to the police and reportedly told them that Ethyl, who had been arrested, was not at fault.

I escorted Frank's body to Louisville, Kentucky where I met his father and two adult sisters. His father was dressed nicely and was well spoken. The sisters, while subdued, were friendly and eager to hear anything I had to say about their beloved brother. The next day I met with a group of American Legion members to plan the military honors that would be rendered during Frank's burial. When the casket was wheeled before the family and me, Frank's sister put her head on my shoulder and wept uncontrollably. It was obvious that Frank's family loved him dearly. After the playing of taps, I presented a folded American flag to the family. Frank's father asked me about the $10,000 the government paid to heirs of deceased military members, but I was not prepared to answer his questions. Ethyl was eventually tried and acquitted based on her plea of self-defense.

After I returned to Florida, I learned that my supervisor was being transferred to Little Rock AFB. His son was in the process of resuming his education after a lengthy hospital stay. My friend Billy was being transferred and my plans to become a permanent part of Elaine's life had been dashed. Although, I could and should have done more to pre-vent her pregnancy, the fact that she didn't want to have children was something that I was unwilling to accept. I told my NCOIC that I would accept the transfer to Little Rock AFB in his place. My offer to transfer to Little Rock AFB was quickly accepted and approved.

Chapter 21
Meaningful Relationships Take Time

Little Rock AFB was about twenty miles from Little Rock, Arkansas. *General* Curtis E. LeMay was regarded as the father of the Strategic Air Command (SAC). He controlled many bases, including Little Rock AFB, from his Offutt AFB headquarters in Omaha, Nebraska. Earlier in his career he had directed an effective but sometimes disputed B-29 bombing campaign in the Pacific theater during World War II. It is estimated that as many as 900,000 Japanese were killed because of his indiscriminate bombing campaigns. He was a man who gave people the impression that he was all powerful. On one occasion when advised that the cigar he was smoking around an aircraft might cause an explosion, his forceful response was, "It wouldn't dare. It seemed to those of us who having once been assigned to a SAC unit, were destined to never leave SAC's command. As a member of SAC, I was sent to Africa on temporary duty in French Morocco. Although I occasionally saw some Africans, for the most part the people I encountered, during those days of widespread French colonization, were French." In 1968 General LeMay was chosen by George Wallace to run as his Vice President. But LeMay's expressed willingness to employ nuclear weapons against Russia and other countries frightened the voters and reportedly under-mined Wallace's presidential bid.

I rented a room in Little Rock from a motherly landlady who sometimes washed my clothing and occasionally cooked for me. Local entertainment was plentiful, and I usually hung out where dancing was permitted. One day I saw the café's cashier seated with some Arkansas Baptist College students. I had wanted to talk to her, but she had always been too busy. When I went over and asked her to dance, she accepted my invitation and told me she had seen me dance before. We danced around the floor as if we were one person. She really understood her role when dancing with a man. The next time I saw her, she was performing her cashier duties. When I asked for a date, she did not say yes or no. Weeks later when she agreed to date me, I had stopped dating other women. This young lady whose attention I was trying to capture was one of four female siblings raised by a single mother. She came from a rural farming community on the outskirts of Hope, Arkansas and was currently a full-time college student in her senior year. She lived in a rented room off campus. Her landlady, Marie, was a widowed lady with a six-year-old daughter named Dell. Markena had mastered the art of saying no without using the word. For example, the place didn't exist, no matter how elaborate, where we could share some private time together. Somehow, it just was not suitable for her. It was obvious that this woman had no intention of sleeping around and even if she did, she was not going to sleep with me. While that was normally a deal breaker, I decided to have an honest conversation with her. I made it clear that I was an enlisted man who planned to retire from the Air Force. I also told her that some airmen who had relationships with educated women too often fell short of the women's expectations. In many cases, educated women could not cook or properly maintain a home and I would never earn enough to hire housemaids. Kena called me and outlined a plan that called for me to give her the money I spent for supper and let her prepare my meals. The day of my first scheduled meal was warm and an unlocked screen door allowed me to move toward the scent of a tasty meal. She was in the kitchen removing a pot from the oven. The meals that week were excellent, and she even returned some of the money, that she had not spent, back to me.

Markena was graduating from Arkansas Baptist College with a degree in elementary education. I looked forward to meeting her mother. As the graduation ceremonies were about to begin, I thought it was a good idea to let her mother, Mrs. Doddie Colbert, know that my intentions regarding Markena were honorable. When I told her that I wanted to marry her daughter, my knight in shining armor approach didn't impress her at all. As a matter of fact, her quick response was, "I'll never speak to her again!" This was not the response I had expected, and I figured the best course of action was to be quiet. Mrs. Colbert, who had been so happy to be reunited with her daughter, was suddenly hearing shocking words from a man she had just met and knew nothing about. She alone had struggled to help pay for her daughter's education. Now she was facing the prospects of continuing to be alone while the last of her four daughters left the homestead. Despite her mother's objections, Markena agreed to marry me. We decided to have our ceremony, with invited guests on June 1, 1956. Her mother did not plan to attend the wedding. But my military unit scheduled a surprise early morning operation that was launched from an isolated staging area. I could not reach Kena to let her know I would not be at our wedding.

When I saw Kena, there was a drink in front of her, which was unusual. Our guest had left, but I persuaded Kena and an angry Marie to drive to Memphis, Tennessee with me. A friend told me I could get a marriage license and be married on the same day in Mississippi. After spending the night in a Memphis hotel, we drove to Hernando's Bed and Breakfast Hideaway in Mississippi. But they did not have a justice of the peace available. Then our luck kind of changed…I think! We stopped at a grocery store run by a hillbilly like couple and asked if they knew where we could be married. The wife left the store and soon returned with a person of color. He had been farming and hadn't changed his clothes. Willie, still sweaty and dressed in overalls, was not only a local pastor but a justice of the peace. I think I can safely say that no girl ever dreamt about being a part of such a wedding, especially if it was her own.

Mrs. Markena Howard

After driving Marie and Dell home, Kena and I drove to Marshall, Texas. She had enrolled in Bishop College classes that would qualify her to teach in both Texas and Arkansas. We found a room in a private home and began our honeymoon. Love can be so satisfying, but newlyweds still must eat. Kena finally got up to prepare breakfast. After a while I also got up and took a shower. I then went into the kitchen only to find Kena with an unopened can in one hand and a can opener in the other. It appeared that she didn't know which end was up. I'd been duped. Marie had prepared those delicious meals. I would have made the biggest mistake of my life if I had left Kena. Less than thirty days after being certified to teach, she studied my meal choices and easily surpassed me in their preparation.

About a year later we attended a squadron-sponsored, family picnic over the Labor Day weekend. More than fifty couples were in attendance, but only one other couple was Black. Kena and I sat with that couple and four more White couples. Kena's dancing as usual was superb and we swapped partners with the Black couple until they left. One of the White ladies asked me to show her and her husband how to do some dance moves that Kena and I had done. Soon all the couples at our table were doing different moves with each other. We left the picnic which was still going strong to attend another event in town.

The first military person I talked to after Labor Day was my squadron commander. The Colonel asked me if I was aware of what had happened after I left the picnic. Since I didn't know what he was referring to, I replied, "No Sir". I later learned that in 1957 Arkansas was in the process of integrating its previously segregated public schools. An extremely vocal opponent of such change was Governor Orval Faubus. Many citizens both military and civilians alike supported his view. As a result of the impending federal action, there were sympathizers on both sides of the issue. Because Kena and I did not have any children, we were unaware of the heightened tensions regarding integration. Shortly after we left, some aircraft maintenance personnel approached the aircrew people at our table and accused them of trying to rush integration by dancing with a 'N'. I was told that the aircrew and opposing maintenance groups formed quickly and a short but intense brawl, that was stopped by our squadron commander, followed.

Chapter 22
Subordinates and Non-Duty Related Issues

I was eagerly awaiting Kena's arrival to Yakota Air Base in Japan. Yakota had a great NCO club, good food and top entertainers. My wife finally arrived, and we rented an off-base apartment that was close to the main gate. We were anxious to learn new and interesting details about this fascinating country. Can you imagine holding hands with a loved one on the imperial palace bridge? We felt a special thrill while also viewing the Tokyo Imperial Palace from such a close and ideal vantage point.

In my job, we primarily used aerial photography to verify the accuracy of the intelligence we had gathered. It was not unusual for us to discover new and startling facts about targets of opportunity. Most airmen pursued female relationships when off duty. Sometimes I saw my men in airmen or NCO clubs along with their dates. Yakota Air Base was five miles from the city of Fussa, and about thirty-five miles east of Tokyo. While there were numerous activities on base, the city of Tachikawa, also about five miles from Yakota, provided most Black airmen ample opportunities for fun, and games when off duty.

Airmen tended to segregate themselves when off-base. Their interests revolved around bars, night clubs and girls. Girls roaming the streets were sometimes dressed as *geisha*s. But *wearing* a red dress and painting their faces white did not make a person a *geisha*. The role of the geisha in Japanese culture was a high-status one. They were educated, artistic and forbidden to sell sex. Girls who held jobs on base generally did not engage in sexual activity for payment. But the main elements of Japanese society that many military men associated with were pros- titutes. Our government tried to discourage such pairings by initially barring the women from entering the United States. Those restrictions were lifted after a while, and the women married to servicemen were allowed to become citizens. I once was approached by a woman in Fussa who called out to me and asked if I wanted to have a good time. She had an unusual blemish that I had seen on the neck of a woman with whom one of my subordinates lived. He was not shy about his feelings and had stated he wanted to marry her. At the ripe old age of twenty-seven,

I had evolved from an unbridled womanizer, into a now married man and concerned supervisor. I could be silent or discuss the incident with my subordinate. Although untrained to be an advisor in matters of the heart, I chose the latter. The day after my novice counselling attempt, the young man respectfully asked if he could speak to me privately. I realized he had talked to his girlfriend, and she had probably denied everything. Afterwards, he shook my hand, and we never discussed the matter again. Counselling, especially when dealing with emotional issues requires formal training.

Our operational and photographic intelligence units worked together on an important project. To celebrate our success, we held a party for all project participants. Multiple snacks were washed down by an overabundance of beer. The visual entertainment was provided by a Japanese businessman who specialized in showing stag movies. The party was well underway, and I also got involved in a beer chug-a-lug that I didn't win. There was much laughter and individuals were staggering around the place as if they might keel over at any minute. The performers in the stag movies were naked and engaging in sex. If anyone was offended, they did not show it. Then the presenter began to show a movie that captured everybody's attention. The party was suddenly turned into a nightmare that revealed the feelings of those in attendance. The woman was shapely, beautiful, naked, blond, and White. She was reclining on a bed as a Black man approached her. Soon they were engaged in sex. However, the excited sexual **ge**stures and loud laughter in the room had ceased. Most of the people had stopped drinking. Others simply left the room. Those previously staggering drunks now appeared completely sober. Their facial expressions reflected their distaste for the stag movie being shown when the only thing that had changed from earlier presentations was the race of the performers. I've used a lot of words to point out that some people are truly unaware of their deeply ingrained true feelings regarding racial issues.

Markena enjoyed Japan and she kept notes about the places she visited. However, a White parent did not want her child to be taught by a Negro. In addition to protesting, she put together a support group to have my wife replaced by a certified White American teacher. However, the principal rejected the groups demands and let it be known he was more than satisfied with Kena's qualifications and her job performance. Despite the ever-present bias and bigotry that Blacks face, we must remain hopeful while doing what we can to peacefully press our demands for limitless equality.

In 1960, I applied for membership in a Masonic fraternity. After a rather painful initiation—even my father would have been impressed—I became a member of Cherry Blossom Lodge No. 9 in Japan. The lodge was under the Jurisdiction of the Most Worshipful Prince Hall Grand Lodge of Massachusetts. After my initiation I became an entered apprentice. Further study led to my becoming **a** fellowcraft mason. And finally, because of even more intense study, I became a master mason.

At Griffiss AFB in New York, my title as NCOIC was meaningless since I was the only enlisted man in my office. My job for the most part dealt with the development of classified briefing reports that were supported by photographic evidence. In the past, the photographic verifying material had been furnished by our photo intelligence units. But in the absence of a photo intelligence unit and the sensitivity of the material, I alone took, developed, and then printed the highly classified films. Learning this skill made me feel closer to my brother, Eddie, who had also worked as a professional photographer. Even though our office was staffed with an abundance of officers there were occasions when I briefed commanders and their staffs. During one such briefings, a colonel asked me a question associated with the word ameliorate. Although I wasn't sure of the words meaning, I was tempted to try and answer his question with what I imagined the word meant. I thought it was just a fancy word for better. But while a good thing can be made better, the subtle difference is that the process of ameliorating a situation is to make something bad or unsatisfactory better. Still any answer that I might have given when I wasn't sure would have been wrong. Information that might influence how a mission could be planned must always be as clear, accurate, uncompromised by careless disclosure, and complete as possible.

Chapter 23
More Labor Day Memories

The year 1960 was event filled, complete with overlapping visits from Juanita, Kena's sister, and five people connected to my family. Juanita lived in the New York City Bronx area and showed up at our home about a week before my family arrived. Kena and I lived in a condominium apartment on Griffiss AFB. It was an up and downstair apartment with a finished basement. We told Juanita to make herself at home and she started moving things around. I didn't want to offend her, because my manhood couldn't endure being knocked out by a woman. A drug addict in the Bronx had attempted to rob Juanita as she entered her apartment. Her friend, Ann, reported that the mugger was happy to see the ambulance that took him to the emergency room after Juanita destroyed him. Kena allowed her sister to make some changes but voiced her displeasure when Juanita went too far.

I had invited my brother, Frank, to visit me during that Labor Day weekend and told him I had arranged for him to have a date with a local resident. When Frank arrived, he brought Albert Mason, my brother-in-law, my niece Patricia, and Eugene Spears, her fiancé. Frank also brought his girlfriend Margaret who was the sister of Walter, my sister Marie's boyfriend. Before I met Kena, I visited Marie and Walter. The visit was enjoyable until suddenly all hell broke loose. Marie and Walter got into a heated argument that became physical. Here was my sister at 105 pounds, if that, being manhandled by a 230-pound man. Even though I wanted to leave, my legs carried me toward the scuffle. Imagine a situation where one of your shoulders is pressed against the center of a burly man's chest, while his ape like arms encircle your other arm as he lifts you off your feet. If you kick backwards or forwards, you will be kicking empty air and your twisted karate kicks lack the force necessary for you to free yourself. Suddenly, you hear your concerned sister's voice. I couldn't believe my ears, but she was begging me not to hurt Walter. My angry response was, "If I ever get loose, I'm going to hurt you!" Walter started laughing and let me go. As I left, I said the next time they wanted to have make up love, they needed to give me five minutes to get out of town. While my wording of the incident might amuse some, I discovered that for far too many people, domestic violence or being battered was an accepted part of life.

On this 1960 Labor Day weekend, Margaret, entered my home and headed straight for a Grundig entertainment console in our living room. While pouring herself a drink that hadn't been offered, she saw German words on the console. She asked Kena what the words meant. When Kena replied that she didn't speak German, Margaret loudly remarked, "She's supposed to be a teacher". As Juanita was rising to her feet, I blocked her path to Margaret. My brother brought Margaret because she had a car and saved him some travelling expenses. Patsy, Eugene, and Mason just jumped in the car for the ride. But I had to accommodate four more unmarried guests who wouldn't be allowed to sleep together in Kena's house.

Sis was the local date I had arranged for my brother. She was light and had long black hair. Her beautiful face, hourglass figure and shapely legs completed the eye pleasing package. After Margaret was sure that Sis was occupied and wouldn't attract Frank, she started flirting with a member of the band. Gene and Patsy were lovingly interacting with each other, and Juanita was flirting with the bar tender who owned the club. Surprisingly, everybody seemed to have had a good time. We returned home and Frank who'd had a little too much to drink, went to sleep in the basement. Just when everyone was resigned to the fact that the day was over, Juanita and Margaret entered the basement. Margaret asked, "Why can't I sleep with Frank?" She then volunteered to reveal information about how my brother slept. She reached down while talking and lifted the blanket covering his naked body. The next thing I saw was Juanita almost lifting Margaret up the stairs while telling her it was time for everybody to go to bed.

Chapter 24
Continental Adventures

I could hardly contain my joy after being assigned to France. I arrived at Toul-Rosieres Air Base and was assigned to the group level intelligence office. Because I was the ranking T/Sgt with the most time in that pay grade, I should have become the group-level intelligence NCOIC. But the officer-in-charge was not about to spend his days with a Black airman blocking his view. So, he kept the White T/Sgt, and I was reassigned to a squadron level position where I would work directly with combat aircrews. and be actively involved in the many aspects of ongoing mission planning. This *blessing in disguise* included the actual plotting of potential routes on maps. Airmen under my supervision had been formally trained in operational intelligence. Their work ethics were above reproach as they eagerly and willingly engaged in meaningful research. My duties allowed me to be involved in the many associated activities that an intelligence technician is expected to learn. In addition to doing original and creative research, I was able to establish the accuracy of that research by using our squadron's flight generated data. I also developed skills associated with topography as it relates to the forms and features of land surfaces and how to use those land formations to avoid visual detection by the enemy. I personally briefed aircrews on relevant evasion and escape matters. There were always advantages we could exploit when we knew the sympathies of the residents in operational area. The photo intelligence laboratory was in a facility adjacent to our work area and their contribution to our overall mission's success was very important. I worked closely with photo intelligence personnel and when we were off duty, the photo shop NCOIC, who was a senior master sergeant (SM/Sgt), and his S/Sgt assistant, helped me. They assisted me in renovating an off-base apartment where Markena and I would live when she arrived. We even cleaned up empty beer bottles, whose contents we had consumed.

On November 23, 1963, I felt uneasy as I went to work because I received unusual attention from French citizens. I soon learned that President Kennedy had been killed in Dallas. Kena and I soon moved to Rosieres-en-Haye, which was closer to our jobs. Kena had been hired to teach school at Chambley Air Base which was about ten miles from Toul Rosieres Airbase. One day, Lieutenant Amar, my intelligence officer, took me aside and told me that Kena was involved in an automobile accident. Although the car rolled over and was severely damaged, Kena miraculously was not injured.

During one vacation, Kena and I slept in Bayonne Ville, a French village near the Spanish border. I tried without success to find out if my hometown of Bayonne had derived its name from this French village. In Madrid, Spain, I asked a citizen for housing directions. He suggested a place near Avenida San Jose Antonio. We took his advice and rented a room in a large establishment. It was a room with an adjacent and very expansive tiled floor toilet and shower area. Our clothing when put outside our door would be waiting for us all cleaned and refreshed the next morning. We enjoyed multi course food choices and the hotel got us choice seats at the Madrid bull fighting arena. Spain was enjoyable, but Rome brings a smile to my face. People get excited as they talk about St. Peter's Basilica, the Sistine Chapel, the Apostolic Palace, the Vatican City Museum, and other sites. Never-the-less, I will focus primarily on France, the country in which I was stationed for four years. I think a good place for me to start is with the Paris Louvre *Museum,* which is the world's largest art *museum.* It's eight-hundred-year-old collection of the most creative treasures that the world is likely to see, included the Mona Lisa. Imagine Paris, displaying its splendor in all directions as you view it from the world-renowned Eiffel Tower. Then cast your eyes on the Arc de Triomphe, located near the banks of the Seine River and at the center of twelve, outwardly radiating circular avenues. That extraordinary exhibit was erected in 1806 to commemorate Emperor Napoleon Bonaparte's victory at Austerlitz. The triumph also known as the Battle of the Three Emperors is generally viewed as Napoleon's best victory. The viewing of the many painstakingly crafted pictorial, and structural master works, was followed by a romantic boat ride with dinner that Kena and I shared on the Seine River as the sun was setting. The next day we enjoyed a night of fun with Kena's sister Ernestine and her husband John Stovall. We spent the evening together in a beautiful café and dined on a tasty meal. John was an upper-level civilian executive in the Air Force's financial retirement center in Denver, Colorado. On May 11, 1970, he received an honorary award from the National Common Service Achievement Organization. Ernestine oversaw activities of the National Association of Fashion and Accessory Designers Incorporated. She was their leader for four years.

The founder of the first Black lodge within the Masonic Order in America was *Prince* Hall (1735–1807). As a colored citizen of Boston in the Revolutionary War era, he pointed to the hypocrisy of a war being waged in the name of freedom by people who enslaved others. His efforts, with the help of others, was responsible for the birth of Liberia. He convinced others to take part in his movements, and he was a spiritual teacher and master craftsman type builder. He was also a landowner, a Revolutionary War veteran, and a legal scholar. Before the Revolutionary War, Prince Hall and fourteen free Black men petitioned for admittance to the White Boston St. John's Lodge and were rejected. North American Grand Lodges denounced Prince Hall Lodges and Prince *Hall Masons*. They viewed them as illegitimate and refused to recognize their legitimacy. Today the few mainstream state Grand Lodges that do not recognize Prince Hall Grand Lodges are, except for *West Virginia*, located in southern states. Prince Hall Freemasonry was founded on September 29, 1784 and consist of two main branches: The independent State Prince Hall Grand Lodges, and those under the jurisdiction of the National Grand Lodge. I heard that our Prince Hall Masonic Fraternity's membership included men like Dr. Martin L. King Jr. Today there are over 300,000 African American Masons.

France itself was engaged in major activities related to Masonry. They were comfortable when visiting our lodge, and we were welcomed in their midst. You have not really dined in France until you have been invited to a dinner in the private home of a French family. I was invited to such a meal along with Kena, Captain Mitchell, the officer-in-charge of my intelligence section, and his date. The elaborate meal, prepared and served by my Masonic brother, was far beyond my expectations. The food was served in so many courses that for all practical purposes the food, in and of itself, was the entertainment.

In 1965 I was installed as the Worshipful Master of Feagins Lodge Number 17, Free & Accepted. Masons, located in Verdun, France. Our Masonic lodge operated under the jurisdiction of the Grand Lodge of Massachusetts. We were invited to an International Ball and Banquet. The affair was hosted by The Grand Orient de France (GODF) in Paris. It was a huge affair attended by ambassadors and dignitaries from Belgium, the Netherlands, Germany, Austria, and nations as far away as the Gold Coast of Africa. The menu included an assortment of international foods. The musical entertainment for the most part was stately and serene. My Masonic brother Francis Williams and his wife, Dorothy were there. She was a tall, statuesque, and beautiful woman who was dressed in a gorgeous gown. As such, she was a picture of dignified American womanhood. I was surprised, in this international setting, to suddenly hear Rufus Thomas' version of Walking the Dog. The song was a very lively version of soul music. But I was more surprised to see Dorothy partially break down toward the floor and start doing her interpretation of the dance. Rather than act shocked, other revelers looked at her only long enough to attempt to copy her movements and do the dance themselves. No matter what other experiences I have in this old world, I will never forget that occasion in Paris where dignitaries didn't just pretend to have fun.

The United States hadn't sent me to France to party, and I worked hard to meet the demands of my job. During an inspection of mission planning procedures used by the 19th Tactical Reconnaissance Squadron, the inspectors praised our aircrew folders and asked us to provide them with a step-by-step written guide on how we constructed the combat mission flimsies used by our aircrews. Our squadron intelligence officer had only recently been assigned to our office, so he had me create the requested document for the inspectors. Later, those procedures became a chapter in Air Force Manual 55-5 which mandated that the procedures be used by other similarly equipped squadrons in the European theater. I received praise for my flimsy folder guidelines from my officers, the aircrews, and the squadron commander. I did not receive favorable comments from our group intelligence officer. As a matter of fact, he rated my annual evaluation in a manner that devaluated me. In his remarks he noted that my intelligence officers lacked the experience to accurately assess my performance. The base Wing Personnel Office agreed with me when I pointed out that he was supposed to evaluate me rather than the officers for whom I worked. A corrected copy of my evaluation deleted his remarks in their entirety.

Later, I was selected to become NCOIC of the 22nd Tactical Reconnaissance Squadron. My new job was a real challenge. The office was in an aircraft hangar. It contained a safe that was filled with outdated classified material, a few desks, and some chairs. My officer-in-charge was a captain on his first overseas assignment. I am so thankful that I had worked earlier with a S/Sgt named Ken who possessed and passed on to me some of his carpentry skills. I begged, borrowed, and did everything but steal needed materials. We used those supplies to build map racks that covered huge unsightly heat and air ducts mounted along the full length of a wall at a height of about five feet. We scavenged 4X8 foot sheets of plywood and plexiglass and used those items along with legs cut from 2X4 feet pieces of lumber to build individual aircrew mission planning tables. As we were engaged in these building projects, we were also destroying outdated classified materials and replacing them with up-to-date data and supplies. We then went into a secured portion of the squadron's operation and aircrew briefing area and erected a sixteen feet-wide structure that would support two four feet by eight feet sliding panels upon which pertinent mission planning data could be posted and updated. We added booths in our own office area that permitted us to have two-way conversations with both our operations areas and our airborne crews. I didn't enjoy the danger associated with running the wiring very high up in the hangar. That skill was a byproduct of a time when I used to build hi-fidelity components in base hobby shops. Working closely with our aircrews in France was the equivalent of attending classes every day. There was so much to learn, and the pilots didn't mind sharing their knowledge with me. I also took advantage of invitations to fly with pilots in T-6 trainer aircraft. I'd gone all out to create a first-class mission preparation area and I was later decorated with the Air Force Commendation Medal for the finished product.

In 1966, I was promoted to master sergeant at Grand Forks AFB. A subordinate, Airman Bussey, rooted his improvement in formal education. He had taken the United States Armed Forces Institute (USAFI) college GED test. He received fifteen semester hours of college credit from the University of Maryland after passing the test. He also took college courses taught by fully certified college professors on base. I was inspired by Bussey's passion to take and pass the GED test. I had previously considered becoming a teacher's aide after my military retirement because my time off would coincide with my wife's vacations. After I passed the test, I aimed higher and decided to become a teacher.

Enlisted men sometimes jokingly referred to themselves as *pee-on* or one so lacking in authority that they were only fit to be urinated upon. Even before I failed the OCS test, I never regretted being an enlisted man. My nephew, Joseph, completed a university ROTC program and

was commissioned as an officer in the Army Reserves. He was the first
direct descendent of my mother to receive such a commission. He once
told me that it felt strange for him to be saluted by much older men. In
that connection, I am reminded of a story told by the son of General
Colin Powell. He said his father witnessed an event where an enlisted
man was repeatedly saluting an officer. The officer, who the enlisted
man had initially failed to salute, ordered the man to salute him 100
times. Armed with those details, General Powell informed the officer to
return each salute that he received because the salute is a sign of mutual
respect. The general's actions implied that while those with less authority do not lead their superiors, those who do lead are not the masters of
their followers. I also remember a Mahatma Gandhi quote embraced
by my nephew Lewis Spears. Lewis was enrolled in Harvard's leadership school and quoted Mahatma Gandhi's saying of, "There goes my
people. I must follow them for I am their leader." Lewis, after attending Harvard has returned to Jersey City to continue his guidance of the
youth organization in the hood with whom he had previously worked
hard to provide positive guidance.

Joseph Borroum Sr.

Markena got a job on base and became extremely close to a
co-worker. I became friends with her husband, Ben, and all of us began
hanging out together. Kena and Catherine (we called her Kitty) were
extremely attractive and related to each other like sisters. When they
entered the NCO club, people would remark, here come those foxes.
The presence of chess players on the base, suited me just fine. I signed
up for a base-wide chess tournament. After concentrating my efforts, I

was able to conquer one opponent after another. This was a day that I wound up in the finals. Much to my surprise Ben was my opponent. He was a chess player that I beat every day of the week. Now, only he stood between me and the base championship. I was overconfident and let him beat me with a Queen's gambit or fool's mate move that I had previously taught him. He became the base chess champion.

On February 22, 1967, my sister Ruth gave birth to Theodore, her eight and final child. On March 22, Patricia gave birth to her third and final child, Lorraine. These children, born exactly one month apart, were viewed as a special blessing, but sometimes wonderful gains are offset by terrible losses. In May of 1967, I requested a leave of absence to visit my family. When I arrived, my mother who now lived in the Jersey City projects was in a Jersey City hospital. I called my boss and was told to take as much time as I needed. He also advised me that I might be sent to Vietnam. While she was happy to see me, my mother appeared sad and avoided talks about the future. She even refused to watch her favorite TV soap opera entitled Stella Dallas. Suddenly, she was moved to a medical facility in Secaucus, New Jersey. I went to see her on June 2, which was my eleventh wedding anniversary. My siblings and I drove to see our mother on 3 June. It was a beautiful day and we laughed as we drove along, especially when Eileen's wig flew out of an open window. We arrived at the hospital, and I put the group out. While parking, I noticed the group being approached by my brother Oscar. I could tell by their reaction that they were receiving some terrible news and without hearing the words I knew that my mother had died. I wanted to leave the vehicle to console and be consoled by my family, but I could not move. Suddenly, I pictured Paul Leath again in tears on Avenue C in Bayonne. I no longer had to wonder if a man could cry as I acted out the answer by crying as I had never cried before. At the funeral, my sister Marie kept saying, "That's not my mother" as if saying it over and over would somehow bring our mother back to us. But it was Ruth Ann who declared she would not let stress control her life anymore. That stance in this period of great sadness, inspired me to mentally tell myself that I wouldn't worry about dying during my pending assignment to Vietnam.

My plane landed at Tan Son Nhut Air Base in September of 1967. If I had arrived earlier in the day, I might have been among several departing airmen who were killed at that same terminal during an earlier enemy rocket attack. Vietnam had been colonized by the French around 1883. The French objective was to quickly acquire Indochina's potential wealth. In 1946, Vietnamese forces increased their efforts to free the country from French exploitation. After years of fighting, the French suffered a devastating defeat at the battle of Dien Bien Phu in 1954. This was a time when post World War II Allied forces hoped to contain Russian expansion. Leading this effort was George Kennon, who was designated the father of US Containment Policy. Many believed that if one nation fell to the communist, its neighbors, like dominoes, would topple and become communist nations. The actual separation of North and South Vietnam at the seventeenth parallel came about because of a Chinese sponsored proposal. Although China was a North Vietnam supporter, they did not want a powerful reunited Vietnam, with yet undetermined North or South Vietnam sympathies, on their southern border. The separation agreement, which had major Chinese input, was reached by the United States, Britain, France, Russia, and China at the 1954 Geneva Convention. It was supported by 186 countries worldwide. The division was to be temporary until nationwide elections could be held to reunite the country under one government.

Ho Chi Minh controlled the North Vietnam Politburo and was favored to win any country-wide unifying election. His popularity rose because he had been given credit for the Diem Bien Phu victory. Initially, there were several groups struggling for control of South Vietnam including the French who after moving their supporters to the South Vietnam had partnered with some corrupt elements there. Eventually South Vietnam was led by Ngo Diem Dinh, an anti-Buddhist, celibate Catholic bachelor, who had once aspired to become a priest. When Diem's forces prevailed in their efforts to control South Vietnam, he was neither willing to take part in any nationwide unifying elections nor include Buddhist elements into his government. Never-the-less, the United States assumed the role of protecting South Vietnam, a move that ignored advice from the French not to get involved. President Dinh was overthrown and assassinated on November 2, 1963, in a coup by his own generals. There were eight regime changes in those troubled days because the generals repeatedly quarreled among themselves.

President Kennedy once stated that *One Man Could Make a big difference.* His statement soon became a reality. As Commander-in-Chief he had planned to quietly withdraw one thousand of the almost seventeen thousand U. S. forces operating in South Vietnam by the end of 1963. The remaining troops were to be withdrawn by the end of 1965, thereby totally ending the U. S. presence in Vietnam. Those planned withdrawals never occurred because of one man. On November 22, 1963, our thirty fifth president, John Fitzgerald Kennedy, was assassinated. Had he lived and withdrawn U. S. troops as planned, it is possible that over 40,000 of the 58,479 Americans killed in Vietnam, might have lived. One man, Lee Harvey Oswald had done more than assassinate a president; he had altered the course of history.

Lyndon B. Johnson, our thirty-sixth President, did not set out to alter JFK's plan. However, a military confrontation involving several North Vietnamese torpedo boats and the USS Maddox, as well as an incident involving the USS Turner Joy, gave Johnson a reason to increase rather than reduce U. S. ground forces. The *Gulf* of *Tonkin Resolution* proposed by President Johnson was overwhelmingly ratified by both houses of Congress. It authorized the president to take any necessary measures to repel and prevent attacks against American forces. President Johnson appointed William Westmoreland to command the. forces conducting the war in Vietnam. But instead of lessening the U. S. presence, Westmoreland's war of attrition approach and his repeated requests for additional troops led to a massive buildup of American forces that topped five hundred thousand personnel.

In 1967, Brigadier General William O. Doyle, Jr. volunteered to serve in Vietnam and took command of all Seventh Air Force (7AF) targeting operations. As the NCOIC of 7AF Operational Intelligence, my office came under his jurisdiction. Oddly, I had met and talked directly to a general one day after completing basic training, but I never met General Doyle who regularly received intelligence briefings prepared by our office. As a matter of fact, during my tour of duty, it was mandated that the officer delivering intelligence briefings had to be a field grade officer with a military rank of major or higher. Major Healey, who wrote my performance evaluations, assumed the role of delivering intelligence briefings to America's very high-ranking Vietnam area commanders.

For years, 7AF Intelligence and other base agencies had been limited by computer programmers to only two data input and retrieval slots on the base's computers which used keypunch cards. Counting numbers ranging from zero-zero (00) to ninety-nine (99) were used to represent the actual names of topics or subtopics. Because of this limitation, huge amounts of related data were filed together. As the NCOIC, who supervised the input and retrieval of intelligence data, I devised and offered an alphabetical solution to the problem. I combined the first ten whole counting numbers, with letters of the alphabet. If you began with 0a to 0z (zero a to zero z) followed by 1a to 1z, 2a to 2z, etc., you would have 260 two space identifiers representing specifically named topics or subtopics. If you then put the letters in front of the counting numbers, you would create an additional 260 two space identifiers for a total of 520. If you then used upper case or capital letters, you'd have another 520 two space identifiers for a total of 1,040 plus the original 99 subject identifiers. Now instead of retrieving a large block of information that had to be rescreened, the specific data needed would be filed under its own two place identifier.

Enemy tunneling under and around Allied facilities was a big problem. Some of the NVN tunneling structures were quite elaborate. Whenever our forces were able to locate tunnels, we sent personnel—who were nicknamed tunnel rats—to enter, disrupt the enemy, and destroy such structures. However, our men often encountered booby traps and other hazards. General Doyle devised a method to deal with the problem and got permission to implement his plan. Years later he talked about it in an unclassified Public Broadcasting System documentary. He had received permission to use B-52 bomber aircraft that could hold up to eighty-four 500-pound bombs with surface penetrating capabilities. Those bombs effectively collapsed tunneling around and under our military encampments even when the bombs did not score a direct hit. The information regarding selection of those bombs and the location of their delivery system aircraft was readily available on base computers. Surprisingly, both the bombs and B-52 aircraft delivery system were launched from Anderson AFB, Guam, my old 1950–1951 duty station.

Les Duan was a powerful member of the North Vietnamese Politburo. His primary objective in the 1968 TET offensive was to seize power in South Vietnam by creating a general uprising and causing the defection or switching to the NVN side of major South Vietnam combat elements. In 1967, border area battles were staged by North Vietnam to lure our forces away from major areas. The attack on Con Tien in September of 1967 would be the first of such distracting battles. In addition to attacks at other sites, the North amassed a force of around seven thousand around the U. S. forces at Dak To. Our U. S. marine base at Khe Sanh held about seven thousand American troops and they were surrounded by thirty thousand North Vietnamese troops. As the enemy deployed their troops in 1967, General Westmoreland sent American Forces to *Dak To* (pronounced toe) to ensure that U. S. forces would not be defeated as the French forces had been defeated at the battle of Diem Binh Phu. Such a defeat was LBJs greatest fear. This however was exactly what North Vietnam wanted him to do. One U. S. commander, *General Frederick C. Wyans*, viewed the North Vietnamese attacks differently. He persuaded General Westmoreland to let him move half of his army from a Cambodian border area to defensive positions closer to Saigon and *Tan Son Nhut Air Base*.

Tan Son Nhut Air Base had a central area mortuary that I walked past twice each day. At first you could see bodies of those who were killed being readied for burial. Later, a wall was built to shield the morbid tasks from prying eyes. But the wall did not hold back the horrible smell of death. Soon after I arrived, American forces were engaged in base beautification projects. Charlie, a name for the Viet Cong, waited until we had almost finished removing unsightly, *but life protecting, coverages.* Then the V. C., unleased massive rocket attacks upon the base. The rockets were fired from extremely mobile rocket launchers that could be moved even before the fired rockets had reached their targets. While the damage from the attack could have been much worse, we still had to do a lot of work to restore protective areas. I received a commendation for my work in building an underground fortified structure where our operational intelligence office could continue to function safety. Our stay in that shelter was short because we were very shortly moved into a newly constructed four-story white building. The lawn around the structure was beautifully manicured and it enhanced the beauty of the site. If the building designers had intended to impress the dignitaries coming to such an impressive structure, they easily surpassed their goal. Sometimes our office briefed General Westmoreland who was the overall Vietnam commander, and General Momeyer, our Base Commander. The new building became a prime target for enemy rocket attacks, but it was not hit during my tour. However, the base exchange or shopping center just across the road from the new building was struck by a rocket that caused several casualties.

Chapter 27
My Thoughtless Comfort

Due to severe overcrowding, some personnel were paid to live off base. I along with my landlord, his family, and another M/Sgt shared a three unit ranch type dwelling. The property was on a corner and faced the main street that ran through town. Across the street and to our left, was a small cemetery. A five-story hotel that dwarfed all other structures within a half mile away from us was just to the right of my landlord's property. That hotel housed many Army and Air Force personnel.

Whenever there were major enemy threats, the main gate would be closed, and all personnel would be required to stay on the base. On January 30, 1968, some threats were noted. I did not want to lay on a narrow cot in a Quonset Hut on my birthday. Although I had earlier decided that I was not going to worry about dying in Vietnam, that mind over matter mentality led me to dismiss threats, somewhat like my chess championship attitude. The next day I hurried off base before the gates closed and carelessly exposed myself to great danger. The moods of the people I passed on the way to my dwelling were festive and I was entertained by numerous displays of fireworks, somewhat like our Fourth of July celebrations. As for my mood, I could not have been more satisfied as I looked forward to being in my efficiency apartment, equipped with a television, stuffed lounging chair, private shower, a well-stocked refrigerator and an *Oh so comfortable full-sized bed*. After a night of very restful sleep, I prepared myself for work on this first day of February, but I was surprised to hear what seemed to be more fireworks. Then, just as I was about to open the hallway door facing the graveyard, my landlord was in the doorway, blocking my exit, and excitedly warning me, that there were Viet Cong everywhere and they were killing Americans. I recalled a recent attack on the side of Tan Son Nhut Air Base that was inhabited primarily by our South Vietnamese military allies. Some Enemy attackers had dug under our security fence before being detected and driven away. But now, I was behind enemy controlled territory, during a monumental battle that came to be known as the Tet Offensive, for control of South Vietnam.

The Tet Offensive of January 30, 1968, consisted of simultaneous attacks by over eighty-five thousand North Vietnamese. The attacks were well timed since Tet was a festive period when South Vietnamese troops were usually on holiday. Airmen who left the hotel early that morning, were unaware of the dangers. Furthermore, they lacked the means to defend themselves since weapons were not issued to Air Force personnel as a matter of area policy. Luckily, many Army personnel with weapons were still in the hotel. They quickly set up a defensive perimeter. Later, my fellow apartment renter and I went to the roof of the hotel. He was armed with an M-14 rifle but he only had nine rounds of ammunitions. We saw dozens of Viet Cong enemy workers opening gravesites and removing boxes of weapons and ammunition. He was getting his rifle ready for an attack on the enemy when I told him our Army forces below had not fired a single shot. We didn't know how many enemy forces were in our vicinity and it would be unwise to encourage retaliatory attacks. Suddenly, as if on cue, a Huey helicopter appeared above the hotel and behind us. We could see the gunner manning his weapon in the helicopter's opened rear door. The grave diggers in the cemetery scattered frantically, but some were struck by bullets.

While viewing the helicopter attack, my mind returned to my childhood. On Saturdays, my mother gave each of us 10¢ and we would go to the Strand theater and stay all day. In the many cowboy movies that I saw, the bad guys when shot appeared to have been thrust backward and off their feet by the impact of the bullets. On this day, three grave diggers had been shot, but there was no jerking of their bodies. They simply stopped in their tracks, collapsed awkwardly to the ground, and lay without any further movement where they fell. Witnessing actual death is a lot less dramatic than the death scenes portrayed in movies, but decidedly more permanent.

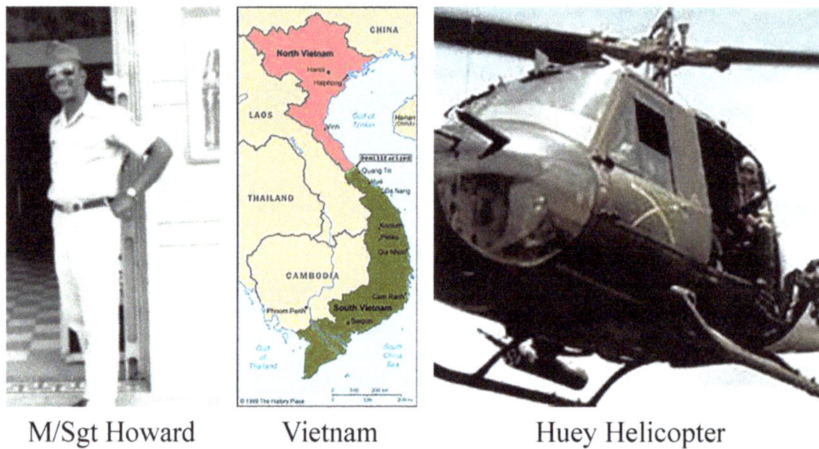

M/Sgt Howard   Vietnam   Huey Helicopter

It was now past midnight as I, fully clothed, laid on my bed and celebrated being alive on my 38th birthday. I could hear the enemy outside, repeatedly shouting, "G. I we're going to kill you." But thanks to the valiant efforts of the Army of the Republic of Vietnam (ARVN), our military police, and our 101st Army airborne forces—We called them the Big Red One—the siege of Saigon and Tan Son Nhut Air Base was soon broken. South Vietnamese military forces did not defect as the invaders had hoped.

The situation in the old Imperial capital of Hue, two hundred miles east of Saigon, was different. Hue had been overrun by the enemy and it took twenty-six days of intense fighting before allied forces could drive the enemy out of the city in bitter house to house battles. The once beautiful city and prior capital of Vietnam, that had been adorned with statues and beautiful artifacts, was nearly destroyed. In contrast to the dire situation in Hue, I was able to safely return to my office on February 3, with one added duty. I was tasked with issuing weapons to myself and other intelligence office personnel. Armed at last in a war zone, I strutted around with a loaded 45 on my hip. I was proud of how our Army and Air Force personnel came together when responding to the common threat. Now I had an even greater sense of patriotism and was determined to be available to do my best to support my comrades *even if I had to miss the comfort of my efficiency apartment and sleep on a cot.*

The damage to base areas occupied by Americans, wasn't as bad as that inflicted upon areas occupied by our Vietnamese forces. It was as if the enemy viewed our South Vietnam allies as traitors for not joining them in battle. The SVN operations building, and residential areas were severely damaged, and unfit for further occupancy. The SVN Base Commander was killed. Even if I had been on base, there wasn't much an unarmed man could have done. If the enemy's primary attacks had been directed toward personnel, instead helicopters and aircraft, it would have been the equivalent of shooting clay pigeons in a shooting gallery. The fighting was so intense that an intelligence officer who had attempted to rescue me and others, was forced to return to the base. Never-the-less, on a day when men risked their lives for me and others, I knew that I, a Black man, was valued. Days later, I watched a slow funeral procession on base. General Nguyen Cao Ky, who served as the Vice President of South Vietnam under President Nguyen Van Thieu, was impeccably dressed in a military uniform. He stood in the rear of the slowly moving open vehicle, bearing the remains of the Vietnamese base commander.

Chapter 28
Health Concerns

I was looking forward to a week of rest and recreation (R&R) in Australia. But I detected a lump that concerned me. I went on sick call and the next thing I knew; I was being flown 200 miles East to the Cam Ranh Bay American hospital. At the hospital, I saw and talked to Sergeant First Class Edwards, a member of my Masonic Lodge. In France he seemed to be a man who would stand in the way of a charging bull. Now his hands shook involuntarily. He had been wounded several times and expressed a concern about being sent to fight again. I also met a White Army private. He had been wounded in his groin area and was being visited by an Army lieutenant. He was clearly in pain as he placed one foot in front of the other as he followed the officer out of sight. He returned to the room seemingly in a state of disbelief and said he was reprimanded for using a captured souvenir weapon to kill the soldier that had wounded him and was about to finish the job. Stories such as these really opened my eyes to the grave dangers and mental stresses others faced that were much greater than mine. I now realized my mind over matter approach of not fearing death didn't consider the lasting emotional stress endured by many of my military brothers. They were engaged in conditions that they brought back home. Reports indicate that an average of about twenty current and former American military men still commit suicide every day.

A decision, right or wrong, regarding the growth in my chest was made and I soon found myself on a plane full of wounded servicemen heading to Clark Air Force Base in the Philippine Islands. One of the evacuees on that plane was a South Korean who was part of the allied force fighting in Vietnam. At the time, protest rallies were widespread in America. Among the things being protested was the use of flamethrowers by allied forces. Our Korean evacuee was horribly burned in battle and is a grim reminder that our enemies also used flamethrowers.

Soon after arriving in the Philippine Islands, I was operated upon. I woke up and learned they had started but stopped the operation. The new plan was to send me to Wilford Hall, a U. S. hospital on Lackland AFB. A doctor had made a chest incision and stopped. I let it be known that I only needed three more days to complete my Vietnam tour of duty. Somebody heard me, because instead of being sent directly to Texas, my return home was routed through Vietnam with a three-day layover. When I returned to the base, I learned my belongings had been collected for shipment back to the United States. My landlord, who had probably saved my life, had switched my refrigerator with a useless one, but the military police made him switch it back. Surprisingly, I learned my masonic brother, F Williams, whose wife had danced to Walking *the Dog* in Paris, was now renting my off-base apartment. Several days later, I was signed into the Wilford Hall Hospital to await the completion of my aborted operation.

Markena had driven to San Antonio and was staying in a local hotel She arrived on base more beautiful than ever. Two days later the operation to remove suspicious tissue had been completed and I was declared cancer free. The next day, I was up and walking around the ward. I also had two options regarding my Air Force career, a medical discharge or reassignment. As soon as I could get a pass from the hospital, I went to Randolph AFB and chose to be assigned to Ramstein AFB in Germany.

Chapter 29
North Atlantic Treaty Organization (NATO)

Our NATO unit had forces from Canada, Germany, and the United States. All I did was sign documents in and out. I expected Kena to visit me during her summer vacation, and had rented a room in Landstuhl, Germany. The family who owned the unit **was a** German couple with a 13-year-old daughter and a 7-year-old son. The daughter treated Kena as if she were family.

Airman Rico, and I were high fidelity enthusiasts. His girlfriend had a six-year-old daughter named Debra. She was an outspoken youngster with an infectious personality and believed there was nothing she did not know. Occasionally, we'd lie on the grassy slopes by the hobby shop, and I wanted to adopt her. In 1944 when I was a child, I had seen Bud Abbott, and Lou Costello, who acted in a "Lost In A Harem movie". I used to imitate one of the characters in the movie who would go wild every time he heard someone say Poco Moco. Debra loved to say Poko Moco and then would run from me in terror when I pretended to be the monster. When Debra's mother learned that I hoped to adopt her, she stopped me from seeing the child while I was in Europe. She loved Debra too much to ever let her go. Years later her mother allowed 13-year-old Debra to vacation with Kena and me in Dallas.

Kena and I went all out to please her. That love was a two-way street and I smiled when I heard Debra talk about Kena's stew which she said knocked her out. Debra's visit coincided with the period when hip hugger hot shorts were popular, and Debra wanted a pair. When I resisted, she told me she would wear them when she returned to Germany, so I bought her a pair. When I looked for Debra the next day, she had already gone outside by herself. I rushed outside to look for her and saw her hanging over a nearby fence in those hot shorts. What disturbed me more than anything else, was the fact that men, who had never spoken to me during the years that I lived in that complex, were staring at her as if she was the featured attraction. I hurriedly took my innocent baby back into the house.

Debra was working at a food concession on base when she met Jim who was White. They married and had two boys. After retiring as a M/Sgt, Jim continued to work as a civilian firefighter while study-ing and graduating from college. His advancement in Colorado as a civilian fireman was both rapid and impressive and he became the top El Paso County Fire Chief in the Colorado area. Jim became a police officer and ran for sheriff. He even taught college classes for years. He is currently the County Director of Emergency Management. Kena and I had stayed with the married couple and their two small boys in California where they lived on base. Joyce, my second wife, and I also stayed with the family in Colorado. One day Hildegarde told Debra, that her biological father, wanted her to meet him and his children. Jim said Deborah told her mother that her father lived in Houston, and she refused his request.

Debra

Jim, **Debra,** and Family

Jeremy & Joshua

Chapter 30
Rising Ambitions

In 1970 my retirement was partially processed overseas, and I was sent to Carswell AFB near Fort Worth, Texas. The first sergeant told me I was to inspect the barracks until my processing had been completed. He lived in the barracks and when I reported violations in his room, he was not pleased. As a matter of fact, he told me to stay in Dallas until my papers were ready to be signed. Before retiring, I enrolled in three correspondence courses because I could not legally apply for them if I were a full-time student elsewhere. I retired on May 31, 1970. I had trouble accepting my growth from that uncharted-era personality to the more decisive man, now on a path to a desirable future.

On June 1, 1970, I entered my first on campus classroom at Bishop College with benefits others could only dream about. My tuition was paid for by the government. My pension covered other bills. Being happily married was so much better than dating and trying to be socially active. As if these advantages were not enough, I could see the campus from my living room window. That first college day was just one day before Kena, and I celebrated our fourteenth wedding anniversary. By the end of my first two summer sessions, I had GED, on base class, and civilian campus credit of 60 semester hours. Prior to the fall semester, I hired a student to tutor me in math. He was the younger brother of a science professor at Bishop College. There were days when I worked after midnight to complete his assignments. Then, seemingly out of nowhere, I had an epiphany or sudden ability to understand and apply upper-level math concepts. Perhaps my ninth-grade math teacher had taught me more than either of us realized. I easily returned to the days of math formulas such as the Pythagorean theorem in which the square of the hypotenuse or longest side of a right triangle is equal to the sum of the squares of the other two sides. The use of letters and symbols to represent numbers and the orders of operations were fully understandable to me. Years later, I taught introductory algebra to some of my elementary students. They were up to the tasks and loved the challenge.

I did well in math, but I really enjoyed history. Because of my military service, I'd been to places that I now read about. I could even add content to some textbooks. My professor, Dr. N. A. Jones was about my age, and he enjoyed our lively exchanges of viewpoints. One day when I entered his class, I found him sitting in my seat. He then announced to the assembled class that Mr. Howard would be teaching history today. We became friends and occasionally he visited my home. He, like Debra, loved Markena's stew. Kena and I also befriended his mother. She pursued and ultimately received her doctorate while in her sixties. from Southern Methodist University in Dallas. Unfortunately, my friend had been undergoing treatment for cancer and, true to his nature, his primary concern was for his mother who he knew he would leave behind. I still miss my mentor and friend, whose mother asked me to be one of his pallbearers after he passed away much too soon.

The dean of education when Bishop College was in Marshall, Texas, held the same position in Dallas. When Kena and I first met him in Marshall, Texas, he put a hand on my knee that I angrily brushed off. However, I let Kena complete the courses she needed to be eligible to teach in Texas and Arkansas. One day, he asked If I would manage some of the younger students he had working in his office. The dean was also the president of the school board in the district where both Kena and I would eventually teach. I told him I would consider managing the students if the job would not keep me from taking the maximum number of classes allowed. My reservations were accepted, and, for the record, he never touched me again. Many students changed their fields of study in midstream. The tendency not only increased their monetary cost but added to the amount of time they had to stay in college before graduating. Since I really didn't know enough about what motivated them to seek changes, I kept my opinions to myself and let a well-qualified dean of education do his job without my novice interference.

One day when I returned home after my classes, I noticed Kena was visibly upset. It appears that her principal had tried to date her. I got on the phone and called the school. The secretary who answered the phone advised me that the principal had already left the premises. I told her I would see him the next day. I had my wife call in sick and I went to the school. He was in his office with his physical education teacher and another male teacher. We didn't have a physical confrontation, but there was lots of shouting and bad language. As I left the school, I heard the P. E. teacher tell the secretary not to call the police. When I arrived at Bishop College, I was met by an angry dean. When he started to criticize me, I told him that my wife had always handled unwanted attention herself, so when she asked me to help her for the first time, I knew her boss had gone way too far. While I had risked being arrested, the problem was quickly resolved. My wife was reassigned to another school within the district the very next day. Her new principal was a very old administrator who later helped me surprise her.

During the summer of 1971, Arthur Ashe visited my Bishop College P. E. class. When I got a chance, I asked him what I could do about a painful condition called tennis elbow. In a humorous response, before the whole class, he told me to stop playing tennis. Later, he made a point of coming to me and offering some rather helpful advice. Arthur Ashe had been a former Army officer at West Point and the first Black man to win Wimbledon, which was the most coveted tennis championship in 1975. He was also, I now realized, a sensitive, and caring human being. Evander Holyfield is another celebrity that I was able to engage in one-on-one conversation. He was standing outside a Las Vegas establishment. He had on coveralls and was very cordial and friendly as he talked to me. At first, I thought it strange for him to be outside and alone at night, but upon reflection I realized one doesn't always require bodyguards when you are the heavy weight boxing champion of the world.

I graduated in December 1971 and was hired to work in Dallas as a teacher's aide at Skyline High School. The school had 3,000 students and first opened its doors in September of 1971. The pupils it served had been transferred from four rival high schools in Dallas. Because of earlier rivalries, the groups of students often clashed with each other, and trouble continuously brewed just beneath the surface.

On my first day in January 1972, I attended a faculty meeting and boldly sat on the front row. Among those in attendance were Superintendent N. Estes, Associate Superintendent B. J. Stamps, Principal E. Davenport, and the entire faculty and staff. Surprisingly, out of this huge body of people, I found myself being approached by Associate Superintendent Stamps. He was the one who told me I would be working as an in-depth counselor. I didn't know what an in-depth counselor did, but I was impressed by the title. My desk was placed in a temporary building, identified as A-12. I soon discovered that students entering A-12 did not do so of their own free will. They were students with discipline problems. I no longer had to figure out what the title of in-depth counselor really meant; I was, in simply stated terms, the room A-12 student jailor. I tried to find books that might help me understand how I could reach these students. I gave countless assignments and lectures in the hope that I could persuade some of the students to act in more socially acceptable ways. I even pointed out how they could personally benefit if their actions improved.

One day, as I was talking to one student, I saw another student out of the corner of my eye who was about to sucker punch me. My response was more a reflex move instead of an avoidance move. It was a karate movement that I had practiced for many years while stationed in Japan. The next thing the attacker knew is that he was on the floor. I had started taking karate lessons in Japan in 1959 alongside an air force karate brown belt holder. That sergeant, Vincent D. worked for me in intelligence. Television in the 1970s was dominated by David Carradine who played the part of Kwai Chang Caine in a television series entitled Kung Fu. Before I could finish eating my lunch, students were asking me if I knew karate. But my thoughts were focused on probably being fired. While I was not fired, it was what the student said that really got my attention. He said I was always talking and using what he called big words that he did not understand. I realized I was so proud to finally finish first at something that I wanted to keep impressing people. It was time for me to return to the classroom and focus on counseling. I needed to learn how to speak with understanding, but still not talk down to my listeners. Bridging the gap, my Bishop College commencement address theme, delivered much later in May took on new meaning. I also discovered behavioral adjustment class guidelines. They were contained in a textbook. The book also suggested that military veterans might be ideally suited to control BAC classes. It now appeared that Associate Superintendent B. J. Stamps had also read that book.

My formal graduation ceremony occurred in May 1972. I had been told by Dean Rollins to prepare and give a speech for the graduating class. The May 6, 1972, edition. of the Dallas Post Tribune printed this story: Bishop College Names Honor Students—Robert Lee Howard, an elementary education major with emphasis in history, will graduate Summa Cum Laude from Bishop College. This is the highest academic honor a student can achieve. Graduating with second honors, Magna Cum Laude, are Helen Benjamin, an English major, James Richard Connor, Accounting, and Lula Pickett, Psychology.

My Commencement Address
Bridging the Gap Between Authoritarianism and Democracy

Good evening. After retiring from the Air Force, I chose a career in public education because it offered me several attractive appeals. Among the appeals was the opportunity to mimic the careers of my brother Oscar and my wife Markena. I would also be able to travel, with Kena during our lengthy summer vacations. While I desired to be part of a respected field, where people would look up to me, the greatest lure was that I've always had a genuine love and affection for children and hoped to help them in their socially acceptable development. Studies here at Bishop College have exposed me to the concepts of authoritarianism and democracy. Reflecting on those terms, made me realize I was starting a career that would be more difficult than I at first imagined. For example, within the context of classroom discipline and my military life, I began to wonder if a person with my background could truly make the transition from authoritarianism to democracy. Many teachers say that children must be subjected to strong discipline to stay focused and on task. That sounded somewhat like military training **to** me.

Military training, however, quickly molds soldiers into personalities that are defined by many as being authoritarian or commanding. The training encourages lock-step patterns of actions and behaviors so that members of fighting teams know what to expect from each other. Those rigid standards of obedience and behavior were tools to increase the chances of survival as well as the attainment of mission goals. Men who questioned or defied their superiors subjected themselves to actions such as written reprimands, reduction in rank, incarceration or even death. Obviously, such methods are inappropriate when working with children. To have values in common, students must be given opportunities to have input into issues affecting their lives. Otherwise, the influences which educate some into masters, educate others into slaves. Plato defined a slave as one who accepts from another the purposes which control his or her conduct. Slavery is found wherever men or women are engaged in an activity that is serviceable, but whose purpose they do not understand and have no personal interest in—the Vietnam War for many protesting Americans was a striking example of this philosophy. While I did everything I could to complete our mission in Vietnam, it appeared to me that the American objectives of stopping the spread of communism and promoting democracy in South Vietnam were not widely understood or supported by our sometimes seemingly complacent South Vietnamese civilians.

The principles of acceptance of democracy create relations that free human personality. Freedom of shared expressions assures that everyone's contributions to the group thinking process will be regarded.

The principle of participation also gains the advantages of pooled intelligence. Students do not build a loyalty to a system when their intelligence is never given an opportunity to shape decisions which have important consequences for their lives. No individual is free who is cut off from his group. Cooperation demands in the final analysis that individuals become free as they achieve ways of sharing without being coerced to do so by others. To deny any of these principles is to reduce the effectiveness of any program aspiring to be democratic. Rightly understood, democracy guards us against undesirable theories of mind, subject matter, and inappropriate action.

As I have attempted to illustrate, the improvement of attitudes and methodologies requires a great deal of conscious thought and practice. It is with this in mind that I pledge to continue my efforts to build the strong bridge that I'll need to share democratic interactions with my students. I must believe that a connecting bridge between the wide gaps of authoritarianism and democracy can be built. Failure is not an option because I would, in essence, be like a derailed train falling off an unfinished crossing. Such a train would drag all the attached cars of children into the endless abyss of low self-esteem, inadequate achievement, and failure. I must continue my education to ensure that I can remain knowledgeable in enabling students to identify, and pursue, a destiny that will satisfy their desires. *They must at some point come to know who they are, what they realistically can achieve, and the difficult self-driven process to which they must commit themselves to succeed.* If I can't bridge the gap of persuading rather than commanding and demanding, then I will be ineffective as a teacher. But there is hope that I'll be successful and continue to succeed because learning for me will be a never-ending process.

Although I was complimented for my presentation, one student asked, "Do you believe students will develop independently or just imitate what they thought you had in your mind?" I did not respond to his comments, but I had to catch myself on several future occasions to avoid returning to some of my past behaviors. I realized that it is not one's ability to finish first that should be honored because a person finishes first each year on every college campus. What should be honored is the willingness to continue making the effort so that desirable acquired knowledge will continue to prosper and grow.

Chapter 31
Post Graduate Challenges

Dr. Nelum, also worked as a professor at a branch of Prairie View A & M University near Dallas. He asked me to enroll in his statistics class at that university. Statistics is the study of numerical information called data that lets researchers look at large set of files and reduce them into meaningful information. The study is usually presented in the form of charts and graphs. Two weeks into the course, Dr. Nelum was hospitalized. He asked me to teach the course as his student assistant. I read the textbook, but it was still Greek to me. I also went over the written lesson plans he had prepared. After he discussed the plans with me, I felt I could at least try to teach the class. He also made tests for all the students, including me, which he graded. Finally, he was released from the hospital and resumed his teaching of the class much to my relief. During this period, I persuaded Kena's principal to help me surprise her. A salesman in a new car followed me to her school and handed her principal the keys to a new Mercury. He then drove her old car away. At the end of the day, the principal watched until Kena was visibly upset before handing her a set of keys. When she wondered what was happening, the principal pointed to the new car and told her to drive her new car home. It was clear to me that she liked my gift because I was smothered with kisses when I came home.

I quit my teacher's aide job at Skyline High School and was hired to work during the upcoming 1972–1973 schoolyear as an elementary teacher in a room next door to Markena's classroom. In June 1972, Kena and I enrolled at East Texas State University in Commerce, Texas. I wasn't going to take a chance on possibly paying a university again to teach myself. We wanted to be challenged, earn our master's degrees, and be certified to teach special education. One of our professors allowed us to work together on our master's thesis which was entitled *Teacher Understanding and Support of the Guidance Program in a Predominantly Black School District.* Much later I saw our original thesis in a Southern Methodist University's library. An SMU student after some simple name changes, had turned it in as their own creation. Unfortunately, that wasn't our only concern. Kena normally sat in the tub when bathing. But, our ETSU apartment only had a shower, and she discovered a lump in her breast while standing. Our joy of getting master's degrees in 1973, was replaced with our breast cancer concerns.

Teachers at Wilmer Elementary School paddled their own children under the watchful eyes of the principal. When I first paddled a child, my upper lip and paddling hand swelled. An alternative method I later used in the upper grades to establish control of my class was a handshake challenge. I discovered it because of shaking hands with Kena's Uncle Wade Van Hook. He was a big man who worked in a factory and farmed a track of his own land. He was extremely proud of his strength, and he squeezed my hand as hard as he could. For some reason, I could feel the pressure, but it was not unbearably painful; maybe it was my bone structure. I later used the knowledge as a class control method. For example, when an opportunity to engage a strong student presented itself, I would allow that student to try and subdue me by squeezing my hand; I never squeezed back. I withstood their best efforts and by doing so was accepted as the unquestioned authority in the class. I only had to get physical three times beyond a handshake. Once at Skyline High School, a second time at Wilmer-Hutchins High School. The third time was at Hogg Middle School when a student entered my room to fight one of my students. I was able to lift him bodily without hurting him and remove him from my room. While I wouldn't recommend these class control methods to others, they worked for me.

One of our particularly troublesome elementary students was George A. Even though he didn't ride the bus, he would hang around and get into trouble after school. As I was about to stop a problem that George was creating, the principal, Mr. Reed, intervened. What Mr. Reed encountered was an out-of-control student who kicked him repeatedly. I'm glad it wasn't me. Years later George was enrolled at Kennedy Curry Middle School while I was there. One day he left the campus without approval, with another White male student who he later reportedly raped at knifepoint. Studies indicate that students who are not successful in their school studies by the third grade, are more likely than not to wind up in prison. George's crime sprees ended when he, as a young adult, was sentenced to prison for life. Police in his hometown reported a significant reduction in area crimes after George's incarceration. The local newspaper asked me to comment on George's scholastic record, but since they changed my name to Tom in the story, I questioned the accuracy of the rest of their report.

Wilmer Elementary, Grade 6. Mr. Ramsey,
Principal & Mr. Howard, Teacher

I had my students display their strengths and weaknesses before each other on the blackboard. After some initial discomfort, the students, as expected, worked harder under revealing conditions to impress their classmates. Thus, their gains were far reaching and as their self-confidence grew, they treated their assignments as games to see who could perform the best.

Kena and I made up for the absence of biological children of our own by teaching our students as if they belonged to us. Our pupils sensed that we cared for them and enjoyed being in our classes. During recess, I allowed the boys to interact with each other and sometimes those fifth and sixth grade boys would try, to get me down. When it came to the girls, I would not allow them to interact physically with the boys in any way. My major classroom rule was that nobody touches our girls not even Mr. Howard. That rule ensured that female students were given the respect they deserved.

Keeping boys and girls from interacting worked well until Rowena entered the picture. She had asked our principal to put her in my class the next schoolyear. I told Mr. Ramsey I didn't want her in my room because she sometimes stopped by my room during class changes and touched the back of my neck. I remembered the Arkansas brawl I had unknowingly caused between the aircrews and the maintenance men when I danced with some of the white women at our Air Force picnic. I wanted to avoid even the appearance of impropriety. Mr. Ramsey must have dismissed my concerns because Rowena was assigned to my class the next schoolyear. So many years later, I'm thankful to him for his decision to assign the child to my class. Children possess and freely spread an incredibly special agape love for those they care about. I was never childless during the years I worked with students like Rowena. She was a respectful, hardworking, and very responsible student who was liked by everyone. It was students like her that let me enjoy the purest and most fulfilling feelings of accomplishments in my role as a teacher

Rowena L

The professor assigned to personally monitor my progress in all phases of administration at ETSU, renamed Texas A&M University-Commerce, was Doctor Stuart Chilton. Because he was very tough, students nicknamed him Chilly Chilton. One day he made a 70-mile visit from ETSU, to observe me. I was about to teach the concept of zero to my math class. At the elementary level, students are taught that zero is used as a placeholder in our place value systems. Zero in higher grades plays a central role as the added identity of numerals, real numbers, and many other algebraic formations. One student, Teresa, had so much difficulty with the concept that I created a homework assignment primarily aimed at helping her. The next day my principal and Doctor Chilton, entered my classroom. After looking at my lesson plans, he abruptly left the room. His sudden departure surprised me, but I had my students solve problems I had placed on the blackboard earlier in the day. Teresa had a challenging place value problem which she not only solved but explained in detail the method that she had used. I was so pleased that I excitedly exclaimed, "I knew you could do it!" At the same time other students in the class began clapping to let Teresa know they were happy for her. It was then that my door which had been partially closed, was fully opened and Doctor Chilton reentered my room. He said it was rare for him to observe a classroom with a climate so favorable for learning.

I was certified as an administrator in August of 1974 and immediately applied for a principal's vacancy at an elementary school. The personnel officer's nephew got the position, so nepotism or the practice of hiring one's own relatives triumphed. My chances of getting the job were slim **anyway** since some of the other job applicants had been trying to become administrators for more than a decade.

Mother Doddie, who I'd grown to love, was living with us. On the occasions when she saw a doctor, I carried her up and down the stairs of our second-floor apartment. Even while ill, she wanted to be in control and I can still visualize her, with pursed lips, whenever she did not get her way. After leaving our home, she went to live with her daughter Ernestine in Denver. Sometimes she would brag to John about how Sonny, as she called me, carried her safely up and down the stairs. John Stovall, Ernestine's husband, was up to the task. He too carried her in his arms to and from the car

Mother Doddie returned to her beloved home on the outskirts of Hope, Arkansas, the city where President Clinton had lived. Kena usually left work before me because I supervised students as they loaded the buses. One day when I returned to our apartment, I saw Kena crying on the couch and learned that Mother Doddie had died. We grieved her loss and people from New York to California paid their respects to this beloved lady. Her funeral was held in the church where she directed the choir for many years. On that sad day you could tell the choir was singing for someone special as they poured their hearts out with uplifting songs. The minister's eulogy paid tribute to one of God's good and faithful servants and she was buried on the church grounds that she had long served with love and affection.

After the loss of Mother Doddie, our apartment was robbed on Thursday and our German entertainment center was damaged. That same weekend, I looked out of my window and saw policemen with shotguns. When I asked a neighbor what was going on, she quickly replied, "You're about to get robbed." She knew we had been robbed before and noticed a strange man coming out of the empty apartment below us. She called Gerard, the maintenance supervisor for our apartment complex. When Gerard appeared, the intruder tried to duck back into the empty apartment below us but was stopped by Gerard who noticed an object in his waistband. Gerard grabbed and removed the object that turned out to be a homemade gun. The police were called and took custody of the man. We thought **we were safe** when the empty apartment under us was rented by a Dallas policewoman. One day as she was about to park her car, an armed man entered her car. As he drove away, she leaped from the moving vehicle. The car thief continued into the heart of Dallas where he was confronted by a uniformed police officer who he shot and killed. Eventually he was captured tried, convicted, and sentenced to death. Kena and I agreed we had overstayed our time in that apartment complex, and it was time to move.

Chapter 32
Career Advancement

Some of my classmates at ETSU were already administrators. On occasions our class was divided into groups and assigned tasks. I realized It wasn't in my best interest to be viewed as the class clown, so I sought ways to contribute to group projects to impress my classmates who could promote me.

The 1976–1977 school year was underway, and I was teaching a fifth-grade class, when Mr. Cole, our district superintendent, entered my classroom. He told me to apply for an assistant principal's job at the high school that was open. Mr. Cole had been my classmate at ETSU. He and I had worked on a group project. Thursday, February 17, 1977, was the first day I served as an assistant principal. It was also the day that Sharon Green, my niece, gave birth to her only child, Shakena Green. The name was formed by using the first three letters of Sharon's name and the last four letters of Markena's name. Sharon obviously thought highly of my wife. While I had worked hard, it seems my promotion was more a matter of who knows you rather than what you know, that was key to my advancement. This was a time when keypunch cards were used to input and retrieve information into the Dallas Region 10 Computer Center. Considering my military experience with computers, I felt right at home in my new job. I was also sent to other schools to help them begin their implementation of computerized record keeping.

The brick high school buildings had a separate gymnasium that also served as an auditorium. There were three assistant principals: B. Jackson, T. C. Crawford, and C. Madrigal. They had served as administrators years before my arrival. In addition to monitoring computer activities, I controlled students during lunch and after school bus dismissal activities. Mr. Campbell, our principal was a no-nonsense disciplinarian. The students not only accepted being paddled but wrote a play that was presented to the rest of the student body. In the play, they pretended to be Mr. Campbell in the act of paddling students as if in a frenzy over here, over there, and just about everywhere. The play was well received and some of the students were still talking and laughing about the play weeks later. After the 1976–1977 schoolyear, Mr. Crawford was transferred and became an assistant principal at Kennedy Curry Middle School.

One day I saw a driver hand something to James, one of our troubled youngsters. James had lifted a male substitute teacher who was blocking the door to keep him inside. Also, an administrator had gotten physical with a student in an auditorium full of students and found himself being manhandled. To avoid a similar confrontation, I asked James and a randomly chosen student to help me move some office furniture. When we got to my office, I told James that I wanted to examine the package he was given. At this point, he could see the principal and coach Houston in the next office and made the decision to get pass me. As indicated before, I began practicing karate while stationed in Japan and had continued to train in the sport by going to karate classes in Duncanville, a city just west of Dallas. My class had just been taught a technique that enabled us to use the weight of an attacker against them. James decided it was easier to run over me than to try to leave through the principal's office. One moment he was attempting to run over me only to find himself being lifted and placed in a seat. Mr. Campbell was kind of strong like his cousin Earl Campbell, the Heisman Trophy winning Texans and Oilers running back. I told Mr., Campbell about the package, and he reached into James' pocket. James resisted and it took both Mr. Houston, our head coach, and me to free Mr. Campbell's arm. Mrs. Randle, our school secretary, had already called Joe Ed, the local Sheriff. The package contained 101 marijuana cigarettes and Joe Ed arrested James for possession. The incident and the story didn't end there. During the parental conference preceding James' return to school, we met his grandmother who was his guardian. Even I was shocked to hear her ask for one of the marijuana cigarettes that Joe Ed had confiscated. Years later James turned his life around and got a job working for the district's maintenance department. When I saw him while he was doing maintenance work on my school campus, he was cordial, friendly, and even joked about our prior encounter at the high school.

Wilmer Hutchins High School had an excellent basketball program. One day I saw an unsupervised ninth-grade student enter our empty gymnasium and begin shooting long distance hoops. He stood out because he was only about five foot seven inches in height. I didn't realize it, but I was, for the first time, seeing a future great basketball player in the person of Anthony Jerome *Spud* Webb. He is well known in the Hutch, as Wilmer-Hutchins ISD was often called. His contributions gave us **a** team that could challenge any local area high school. During a 1986 hometown visit, I proudly escorted him around Kennedy-Curry Junior High, where he had previously played, on his way to everlasting fame and glory. He had won the NBA slam dunk title over other great contestants, like Dominic Wilkins who was more than a foot taller than he was. Also outstanding was Mookie, whose brother-in-law was regarded as the Texas heavyweight boxing champion. He told Mookie during a school disciplinary conference that Mookie could accept punishment from one of the principals or be punished by him. Years later, Mookie, served as a teacher at my school. Then there was Rickie Grace, his mother and my sister Eileen Mason worked at the Veterans Administration Hospital in Dallas. Rickie 'Amazing' Grace went on in college to surpass the worthy achievements of his other teammates. In 1988, his last year at Oklahoma, he was a vital part of the team reaching the championship game. He later embarked on an international career and played professional basketball for many years in Australia.

In 1978 our Naval Junior Reserve Officer Training Corp (NJROTC) Commander, my wife Kena and **I**, chaperoned some of our students to a San Diego California Naval base. The government had funded the training trip which included other school districts in the Dallas area. Our students handled themselves with dignity and did well. During that trip, Kena and I travelled **to** Mexico and shopped in Tijuana. I saw chess sets that Neiman Marcus sold for $750. The sets consisted of carved green malachite gemstones pieces resembling Aztec warriors. The opposing Spanish images were carved out of white Topaz or Moonstone and represented a group of conquering Conquistadors. The vertical sides of the chessboard were adorned with more malachite-colored carved figures. After much haggling, among themselves, a competing merchant sold the set to me for only thirty-five dollars.

Flying To San Diego, CA.	Hutch H. S. Students	The Set Decorates My Home

I enjoyed working at the high school, but I missed my classroom students. I still enjoy reading some of their notes in which they indicated they felt good about themselves because of me.

Dear Mr. Howard,

This is the best year of school I have ever had. Math has always been my worst subject until this year when it was my best subject. I can't figure out why I like math so much this year. I guess it's your patience and understanding. Well, if you won't take the quarter, I tried to give you, maybe you will accept my thanks and appreciation.

Yours, Stephanie (Stevie)

Hello Ms. Crowe,

My name is Andrea A. C. I am a former student who was at Hogg Middle School in 1988. Hogg had the most incredible teacher named

Mr. Howard and he taught Special Ed. This wonderful man had an enor-mous effect on my life. There is not a day that goes by that I do not thank God for this man. I believe, all I have completed in this life is due to him. I am writing to you in hope for guidance to a path which will lead me to my Angel. I would like to tell him how grateful I am. If you can give me any information, I will be very thankful. You can contact me at…

Note: I contacted Andrea and relived some pleasant memories with **a** successful college student.

Mr. Howard

You are well appreciated by me and many others. I have learned that if you get along with your teachers, school is more enjoyable. This year has convinced me to go as far as I can in this old world and be the best. Thanks for SETTING AN EXAMPLE.

Your student and friend: Ray R

Ray gave me an Edgar A. Guest poem, entitled I'd Rather See a Sermon. It can be seen on the internet.

In 1978, Kena suffered a setback in her battle with cancer. She had been receiving treatments since 1973 and we were desperately seeking help from any source. During our summer vacation that year, I took her to Ohio where Ernest W. Angley, an American Christian evangelist and faith healer was based. I went through the faith healing ritual without feeling anything. But the change in Kena was uplifting. It had been a long time since she had been as active as she was on the rest of our drive to New Jersey. Her mood remained upbeat throughout our vacation and for many months after we returned to Texas. For her, there was no ques-tion, that the power of positive thinking yielded positive benefits.

Chapter 33
The Social Climate and Forces of Nature

In 1979 Kena and I became the owners of our very first home. It was a four-bedroom home with one and a half bathrooms. The second-floor passageway between rooms allowed upstairs guest to look down into a living room type den. The den was equipped with a fireplace, and I had a romantic skylight built into the slanted roof covering the den. I also built a covered patio next to our two-car garage. Kena's happiness at owning our home was obvious. She must have imagined just how our home should look. We found a dining room set with an oval tabletop, but Kena searched until she found the set with an oblong tabletop in a store over 200 miles away and had it shipped to our new home near Dallas, Texas.

A topic at the first civic association meeting we attended, dealt with removing *Blacks from the town*. Over forty years later, I still wonder how anyone, while Blacks were in attendance, could be so insensitive. That's the type of mentality that could readily support a lynching. Since the 1850s, Blacks have been lynched at a rate exceeding that of Whites by a margin of about three to one. During Ulysses S. Grant's presidency, lynchings were viewed as acts of murder. Not all lynching's were recorded during the period from 1882–1968. But of the 4,743 that were counted, 73% were black. Emmett Till, a 14-year-old child who was tortured beyond reason is included in that number. The White woman involved in that murder later admitted that she lied during the trial. Why wasn't there an uproar in White America and punishment for slave owners who forcibly raped thousands of slaves. Why have so many Americans accepted and tolerated these misdeeds. Even today the evidence of those misdeeds can be seen in the skin complexions displayed by the descendants of slaves. I am not opposed to consensual relationships that Frank, my brother, enjoyed with his wife of more than fifty years. She died of Covid-19 in 2021.

Frank & Veronica
Howard

Rose M. Howard

One morning as I was preparing to go to work, I noticed that my skylight had been blown off its settings during an overnight storm. I climbed up on the roof, set the skylight back into its mounting, and went to work. Our high school, and the area around it, was in what had come to be known as tornado alley. A tornado is a rapidly rotating circle of air that touches both the Earth and the base of clouds at the same time. This was a day of heavy rains and menacing dark clouds that frightened some students who anticipated great danger. One student even hyperventilated in the principal's office that morning. As the school day ended, I began loading the buses. Suddenly, I noticed an approaching twister, so I quickly unloaded the bus and ushered our students into a center hallway in the building. When the dark columns disappeared, I reloaded the buses. However, the funnel clouds reappeared, and I moved our students into the inner hallway again. Some students panicked and I had to run up the stairs to return them to the first floor. One student then began to hyperventilate, and I had her breathe into a paper bag. As I was helping that student, Mr. Campbell and Mr. Simpson, our band director, passed through the hallway and then left. I told the custodians to let the doors remain open because I wanted the high pressure within the hallways to escape into the tornado created vacuum that formed around the outside of the building. The much higher air pressure inside a closed area has the power to move a building's outer walls into the very low-pressure vacuum and thereby cause a collapse. The windows in my truck which had been blown out were proof of that theory. Although portions of the building collapsed, the area where I kept the doors opened remained intact. Overall, the high school was damaged beyond repair.

Bishop College offered to assist our school district. I was deeply involved in plans to complete the schoolyear. Mrs. Randle, worked feverishly to type my plan for presentation at a district meeting that was already in progress. Central office officials, Mr. Campbell, the high school staff, and students were already seated in the Bishop College auditorium when I gave Mr. Campbell the plan for his signature and approval. The program began with the introduction of central office and high school officials. When my name was called, the students rose and gave me a standing ovation. Although I missed some of my Texas A & M University-Commerce classes, my professor gave me school credit for a copy of my district recovery plan. He wanted to examine how our administration recovered from **a** real-life crisis. On May 12,

1979, I received my second master's degree which was in educational administration.

Chapter 34
The Sensible Choice

While Kena and I were taking a summer course at Southwest Texas State University in 1979, I got a call urging me to see Dr. Lewis, our superintendent. When I returned to Dallas, he told me I would be the Acting Wilmer Hutchins High School Principal. After momentarily hesitating, I accepted his offer and set up a meeting with Mr. Jackson, Miss Madrigal, and Mr. Houston. I had chosen Mr. Houston, our head coach' to be an administrator. Then I went to the home of Mr. Campbell to find out what was going on. He was as much in the dark as I was. Suddenly, what was thought to be a settled issue, was not settled at all. Dr. Lewis suddenly left the district and Mr. Campbell was reinstated as principal. Although he asked me to remain at the high school, I sought a transfer. The principal at Kennedy Curry welcomed me as an assistant principal. But soon after my transfer, he became the district's personnel director. Mr. Crawford became the acting Kennedy Curry principal, and I assisted him.

Kennedy Curry had about eleven hundred students and seventy faculty and staff members. One of my goals at the middle school was to reduce student pregnancies. During that 1979 to 1980 school year, I suddenly developed a heightened memory capability that was both useful and frightening. Those savant-like capabilities were sometimes associated with a neural disorder like autism. In my case, I was suddenly able to memorize the names of all the students on the campus. This capability was useful in that if I couldn't identify someone by name, they didn't belong on school grounds. I carried a camera and took pictures of any intruders on the campus. It really discouraged drug dealers.

One day I saw two boys chasing another boy who they caught up with just across the street from school property. One pursuer began striking the boy with a tire iron. I knew Guy, from my time at the high school. He was at the mercy of Rocky. The third boy was Rocky's brother. In defiance of my principal's order, I left the school grounds and crossed the street. Rocky recognized me and stopped his attack. As we talked, Guy got to his feet and ran away. Since the brothers did not chase Guy,

I returned to my campus. I had disobeyed my boss, but I feared Guy would have been killed and I couldn't just watch and let that happen. Months later, Rocky's father, an ex-convict, was shot to death. Rocky had gone home, shot his sleeping father, and returned to the school as if nothing had happened. He was later convicted of the murder. At the end of the 1978–1979 schoolyear, Mr. Crawford took charge of the district's alternative school program, and I became the principal of Kennedy-Curry Middle School.

The Principal

Kennedy-Curry Middle School

Chapter 35
Unbearable Tragedy

Sadly, my career promotion was preceded **by** a downturn in Kena's health. She had a preview of what to expect. Before I left for Vietnam, we rented a room in a home that was owned by Mrs. Brown and her daughter, Jo Ella. This arrangement came about because I had a craving for some soul food. As we were being served, I asked the manager if he could tell us where to find a long-term safe motel. He said his mother and sister had a vacant room in their home that could be rented. It proved to be an ideal arrangement and was almost like having Kena living with relatives. Months later Mrs. Brown discovered she had cancer and Kena played an active role in Mrs. Brown's homebound care before her death.

As Kena's health deteriorated, we practiced being celibate. However, just laying side by side and holding one another provided us with a level of intimacy that was very satisfying. Rarely did a night pass when we did not hug, kiss, and tell each other of our great love for each other. What I came to regard as our tender intimacy, was elevated to a height that few loving couples ever discover or enjoy. In 1980 Mae

Ethel demonstrated her sisterly love when she left her California home to provide skilled care for Kena in Texas. I wouldn't have been able to do my job if Mae had not been there supporting me. Love reveals itself in several ways. Agape love is a pure, willful, and even sacrificial love where we desire the best for others. Family members still love each other even though the object of such love may on occasion exhibit undesirable traits. Sexual love participants are often physically engaged with one another in an instinctive act to produce children. Agape, phileo, or brotherly love is always present in friendships where people without any expectations of a reward or a gift just try to make others happy.

One day Kena removed a green dress from her closet and told me it was one she liked. Then she put it back in the closet. A few days later when I came home from work, she was not there. Mae Ethyl said that Kena told her it was time for her to go to the hospital. She also said Kena had paused and gazed at her dining room set with its oblong shaped tabletop. She then left the house and did not look back. Kena's three sisters, my sister Eileen, and I were at her bedside when she passed. My sister extended her hand and closed Kena's eyes. The next day, I showed Ernestine the dress Kena had pointed out to me. Stine discovered a slight tear in the dress, and I watched as she carefully displayed her sewing mastery by restoring the dress to a like new condition. In the meantime, Ernestine's husband, John had arrived, followed by my brother Frank. The next morning, Frank sat with me on a couch, and did what I had wanted to do for Paul Sr. He wrapped his arms around my shoulders and comforted me. The schools closed early that December 1980 day, so Markena's coworkers could attend her funeral. Mae Ethyl and I paused to view the body, and then took our seats. Before the services began, I returned to the casket and placed Kena's hand in mine and prayed for her. In 1981, I saw the movie Somewhere in Time. In the picture, Christopher Reeves gave up his present-day life to return to a woman he had loved in the distant past. I loved the music created for the movie by John Barry and really felt a connection to the movie's theme. Someday, I too will be spiritually reunited with all my loved ones in God's kingdom for eternity.

Mae Ethyl, Aunt Stella, Stine   **Ernestine and Juanita**   **Dwight—Mae's Grandson**

After Kena died, I worked even harder to improve our educational programs. J. Boykins and D. Lewis, my assistant principals, contributed significantly to that objective by creating written educational guidelines. I was also anxious to continue to reduce romantic interactions among our students because I had seen too many girls that had been pregnant when they arrived at the high school. My anti-intermingling policies were intended to discourage romantic displays of affection including holding hands. Sometimes schedules were changed to reduce student interaction. Students who realized what was being done, complained to their parents. When the parents did call, the words thank you were almost always a part of our conversations.

However, Sarah violated school policy and I, in a district that sanctioned corporal punishment, paddled her. I had tried to have a conference with her mother without success. One day, after being paddled, she thanked me. I was shocked, but when she again violated school policy, I had a female office worker paddle the child. There was no thank you involved this time as she repeatedly yelled and screamed, "Don't touch me!" Still, I was encouraged after that incident and pleased to note her violations had stopped. Just as I began to think maybe paddling does work, Sarah appeared at the door to my office and placed an unused sanitary napkin on the floor and announced she was finished. I had known teachers who became romantically involved with their principals, but this was a child. Never in my wildest dreams did it ever occur to me that any child would be so outlandish! Had she forgotten that this grandfatherly old man, at least in my own mind, held an exalted position here? *I am the principal!*

That evening a female district worker and I visited Sarah's home. I thought the mother would be upset by her daughter's suggestive actions, but I was wrong. She listened quietly, then picked up the napkin and carried it out of the room. She did not return! After that visit a teacher told me the child might fail. I advised the teacher to keep notes on Sarah, that would help her remember to use different strategies when she taught the child again the next school year. I wasn't really surprised when Sarah's grades improved, and she was promoted to the high school.

Shortly after the start of the new school term, Mr. Campbell, called me on the phone and asked me in a very agitated voice, "Howard, what did you send me?" The cause of his displeasure was, of course, my Sarah. She had gotten into trouble in one of her classes and was sent to his office. While waiting to talk to the principal, both she and a male student, who was also in trouble, left the school with the intention of getting a hotel room. The hotel clerk noting their youth, called a nearby policeman. The police contacted the district truant officer, who returned the students to the high school campus. It was never my intention to punish my old boss, but I found myself grabbing at straws and while I had correctly predicted that Sarah would be someone's worst nightmare, I was desperate to keep her from continuing to be mine. If this child's momentary infatuation were to continue, it may have caused some people to think I was interacting inappropriately with my students. Still, it was wrong for me to tell the child's teacher that she would have to deal with Sarah another year. I suspected that she, like my ninth-grade algebra teacher, would find a way to keep that from happening. I needed to step down from my self-declared exalted position and practice a better leadership style. **At** the very least, I should have started an official file and discussed the matter with my school district supervisor.

Chapter 36
The Matchmaker

I was working late one night in September 1985 when the outside door was opened. I investigated to find out why the alarm had not sounded and discovered Mr. Stansel, the maintenance director, entering the building. He was with Dr. Matthews, our superintendent, who was making an unannounced nighttime inspection. Dr. Matthews later sat in my office and discussed his observations. He also took a letter from his pocket and set it on my desk. It was a *to whom it may concern* handwritten letter that contained a woman's picture. The woman, named Betty Joyce, worked in the Houston school district where he had been the deputy superintendent. She had taught two of his daughters music. Eventually I called the woman and learned she had a teenage daughter named Sonja. She shared custody of Sonja with an ex-husband from whom she had been divorced about fifteen years. Although our calls and written exchanges increased our knowledge of each other, many months went by before we met.

I'd had several romantic involvements during the five-year period after Kena died. One was with a member of my church. I asked a school board member and fellow parishioner to find out if the lady was in a relationship. After she denied being involved, we began dating and were even engaged to be married. One Saturday as we talked on the phone, she answered her door. When she returned, she abruptly ended our conversation. Her change of focus caused me to be suspicious. I drove to her place and parked out of sight. I watched her leave her apartment with a man and get into his car. I didn't try to follow them; I just remained parked and watchful. Hours later, I saw his car return and watched them embrace. She then returned to her apartment alone. I knew I should have been really upset, but I was not. It was kind of how I felt when I discovered that Tommie, my Creole crush, was married and Elaine had aborted our child. When I met her after church the next day, I pulled down the collar of her turtleneck sweater. Her neck was covered with passionate kiss marks. I guess you could say she didn't like the way I told her we were through because she threw my ring at me and showed up at church with her boyfriend a week later.

Gloria was a technician who worked for my dentist. I wanted to have the student activity fund audited. When I mentioned that the student activity fund needed to be audited, Gloria volunteered and discovered discrepancies. The business office agreed with her and corrected the account. We later parted on friendly terms, but my relationship with Yolanda and Cedric Haggerty, her children, continued. I love the

Christmas, Father's Day, and even Easter cards I still get from my voluntary family.

Yolanda, Gloria, & Cedric

Gloria's daughter, Yolanda, or Nesie, was an excellent student. North Texas State University was an early admissions mathematics and science school in Denton, Texas for exceptional high school students from across the state. She graduated from the school and became a registered nurse. Through the years she displayed the talent and leadership ability to manage other nurses. Her brother Cedric emerged from the Dallas Independent School District's talented and gifted program with a four-year scholarship to the prestigious University of Texas (UT). The brain train was still on track when Cedric's daughter Amber was born. She was a meticulous planner and her success, boosted by her business insight and the support of her father, was guaranteed. Nesie's daughter, Alexis, displayed determination and would never let a physical limitation deter her from proving that anything others could do, she could do better. She received her master's degree from the University of Houston in December 2021. Taylor, Nesie's youngest child, let her school achievements attest to her brilliance. She earned an associate college degree while still a high school senior and like her uncle Cedric, she too was the recipient of an academic scholarship to the University of Texas and graduated in May of 2022.

Chapter 37
Alternative Endeavors

Joyce and I finally met in April 1986 when she drove to Dallas. She was attractive and somehow seemed to be familiar. Holding her in my arms felt good as we kissed for the first time. Mr. McBay, my retiring teacher, who was about to open a restaurant, prepared and catered a delicious meal for us in my dining room. When our relationship improved, I asked her to marry me, and she said yes.

Her Houston family at our wedding included her father, Hiram Jackson, Sr., her brother John Roderick Jackson, and her daughter Sonja R. Stevenson. My family included Louise Allen, Eileen Mason, Sharon Green, Shakena Green, Lynnetta Young, Patricia Spears, Ernestine Stovall, John Stovall, Kenneth Borroum, Charles Howard, and Paul Larry Leath, Jr. About eighty-five additional friends were there including my Assistant Principal Billy Jackson, as well as my school secretary Ann and Wes Booth, her husband. On August 10, 1986, Joyce and I were married. My big kid Charlz was our Disc Jockey, while Kenneth Borroum and Paul Larry Leath took great wedding pictures. The Groovy Grill, a homey type of restaurant and cafe in Houston only charged us for the food. Each guest had their choice of seafood, beef, or chicken, plus side dishes and desert. If you exclude the rings, the wedding cost less than a thousand dollars but was spectacular when compared to my first wedding.

Mrs. Betty J. Howard

My wedding photographer nephew, Kenneth Borroum, had applied his Air Force skills as a communication specialist to civilian life. He rose rapidly in his Communication (MCI) job to supervise company subordinates. The other nephew, Paul Larry Leath Jr. graduated from Howard University. Today he owns and operates a multi vehicle limousine service in South New Jersey.

As a military veteran in Texas, I was awarded five additional years of teacher retirement credit. However, I had only worked fifteen years in education and was still below sixty. As a result, my teacher's retirement pension was drastically reduced. I previously had success in buying and selling auctioned cars in Dallas. I wanted to expand that type of business in Houston by buying auctioned vehicles and placing them for sale on several used car lots owned by other dealers. The lot owners would receive a predetermined share of my profits. But I learned that if the owner of any lot failed to pay their taxes, then I would be regarded by the IRS as a co-owner and held responsible for all delinquent taxes. While most lot owners assured me, I would not assume their tax burden, they wouldn't sign a tax waiver to that effect. Therefore, I chose not to pursue a business that could have ruined me financially. I repaid all retirement payments and applied to resume my teaching career in my wife's hometown of Houston.

Sonja, my 16-year-old potential stepdaughter had told me it would be okay if I married her mother. But she didn't want anyone marrying either of her parents, even though her father had already remarried. One morning, after my marriage, I discovered Sonja had badly scorched Kena's oblong table in a hateful act. Joyce punished Sonja and handed the belt to me as she left the room. While I was angry, I knew that the child's emotional needs would require much more than a paddling and I did not punish her. After her father's death, Sonja seemed proud to tell people I was her daddy and that she loved me. Sonja began having children and even though we did not live together, I was a primary male figure in the children's lives. Years later, when the near six-foot males were on the verge of manhood, I talked to one of the boys about some missing articles. I guess I was taking too long because the next thing I saw was an open palm two inches away from my nose. Suddenly, that military enforcer was back; but now my self-control was intact. I didn't punch or kick this young man that I had grown to love. However, I didn't want to talk to his hand either. When I grabbed him, his strong resistance told me that I may have bitten off more than a man in his eighties should have done. Still, I was determined not to lose and eventually lifted him off his feet and pinned him to the ground. When he demanded in his best bass voice that I let him go, I told him that a real man shouldn't have to ask another man to let him go. A week later, Sydney, who had watched and was now six, told me her brother had been mean to her. I may have injured her brother's pride, but I strained myself during the lift. That led to a tear and an infection of the pouches in my colon. Diverticulitis is the name for such a hernia like condition. Years later I foolishly lifted a heavy safe to dislodge a clip in the door. That act hospitalized me for a week because of blood in my stool.

Sonja, Jacobi, Jabari, Jarvis & Sydney—2007

After serving in the Air Force, Joyce's brother, Hiram Jackson Jr., became an electrical engineer and later worked for the CIA. In 1987 during a visit to his home in Virginia bordering the District of Columbia, I experienced a surprising flashback. It occurred at the Vietnam War Memorial. As I approached the place where the two walls came together, I was struck by the overpowering smell of death. It was a Vietnam flashback, associated with my daily passing of the mortuary area. I was tempted to pinch Joyce to make sure I wasn't dreaming but her reaction probably would have led to my name being added to the wall. On a serious note, the event gave me a limited sense of how PTSD effected our armed combatants and people who were repeatedly exposed to excessive stress.

Joyce had earned a master's degree in education with emphasis in music. Her passion for music was evident when she voluntarily arranged to have her music students perform, year after year, in Houston shopping malls to lift the spirits of holiday shoppers. She also sang in the United Methodist Church choir. I remember an occasion when I never felt more welcome than the time when Attorney Michael Craig, left his seat among the congregation and escorted me to an isle where he and his wife, Dr. Betty Divine, the retired choir director, were seated. Joyce, who we all heard sing in the choir that day, eventually created her own *God's Amazing Love* album.

Her brother Roderick Jackson possessed extraordinary cooking skills. No one was surprised when he acquired and held the position of head chef at a nationally famous Sheraton Hotel in Houston.

Sandra, Joyce's sister, aspired to be a lawyer and worked hard to attain her challenging goal. After previously practicing law for many years in Georgia she eventually set up her own law firm in Houston. Now retired she has returned to Georgia where her son, William, and his family live.

Deborah and her Nigerian husband, Maurice Ekwo, both earned Educational Doctorates and moved to Africa where she served as the head of the Nigerian Cross River University's Fine Arts Department for 17 years. Professor Maurice Ekwo currently teaches at Texas Southern University in Houston. After reestablishing her residency back in America, Deborah worked for the Houston Independent School District as a teacher and a math consultant. Education, for her seems to be a never-ending process and she graduated again in 2020 with a degree in theology.

Gloria, like her siblings was motivated to be economically independent. She attended and completed business school. She initially worked for the Post Office. After leaving the post office, she has been employed by the Houston Independent school District in various positions for many years. The fact that all the Jackson children became both educated and economically independent supports the theory that the presence *in the home* of a strong and motivating father figure is an irreplaceable asset.

The Jacksons—Hiram Sr., Hiram Jr., Joyce, Rod, Sandra, Debra, Gloria

My application for reinstatement was approved and I was hired by the Houston ISD to work at Hogg Middle School. Mrs. Hill, the principal, had me teach special education math classes. Our head teacher was Mr. Calhoun. Our sixth to eighth grade students were predominantly Hispanic. The students assigned to me were educable and did not need close supervision. The new BAC teacher quickly learned that he would have difficulty managing his class. One student threw a book at his head on day one. He resigned after two months. My readjustment to the classroom was less traumatic. However, I still demanded that my female students be respected. I beamed with pride when I heard my boys tell a new student that 'nobody touches Mr. Howard's girls, not even Mr. Howard.

For the most part organized, and disruptive gang conflicts took place away from the school. That changed one day when students were leaving the campus and going home. Two student leaders of rival gangs had confronted each other in one of the classes. Somehow, one gang leaders found himself alone at the end of the school day. He was being confronted by another leader and a sizeable number of that rival's gang members. It just so happened that Mr. Hogan and I, were assigned to monitor that area when school was dismissed. We called the office to tell them what was about to happen and learned the principal was not on campus and the office workers were afraid to call the police. As the gang began to move toward the isolated gang leader, Mr. Hogan and I put the threatened student between us to protect him. Suddenly, a grown man riding a bicycle towards us from off campus, leaped off his still moving bicycle, with a weapon to attack us. Just as quickly, several members of the threatening gang surrounded him and kept him from attacking the three of us. Because Mr. Hogan and I had previously earned the respect of some of the gang members they let us protect and escort the student to the office.

Theodore is the son of my sister Ruth and Leon Borroum, Sr., her Navy veteran husband. Ted always worked hard in school and completed college with Summa Cum Laude honors. He worked for Deutsche Bank and was scheduled to attend a seminar in one of the nearby twin towers on 9/11. Luckily, he got stuck in traffic and missed the seminar. Also, the Deutsche Bank, where he normally worked, was damaged when twin towers debris fell on the bank. Our family believes God saved him. At Morgan Stanley, his next employer, he dealt with a variety of financial services that offered everything from advising clients on mergers and acquisitions to raising capital for corporations. Later, he worked for Dun & Bradstreet which helps companies improve their performance. As a global leader in commercial data and analytics, their acquired insight from data enabled their clients to compete, grow and thrive. Firms of all sizes rely on Dun & Bradstreet's data, insight, and analytics. But Morgan Stanley, valued Ted's skills and enticed him to return to their company by promoting him and raising his salary.

Chantal, a very opinionated grandchild, decided to become a lawyer. It appeared to be an impossible goal due to monetary limitations. She, however, was determined not to let anything stand in her way. She proved that she was up to the challenge and after passing the bar exam, she is now working as an attorney who is licensed to practice law in the states of both New Jersey and New York.

Regina is a mother that Lewis adores. Although raised in the Jersey City hood, he inscribed the copy of his book that he sent me as follows: I am because you are. Thank you. I love you. He became a teacher and an educational dean of instruction. The group he created continues to be a positive force within a Jersey City neighborhood to which he has returned, after attending Harvard University.

The extraordinary achievements of these family members came about because of their drive. The only thing they needed was the unimpeded opportunity to become all that they desired to be.

| Lewis W. Spears— From Hood To Harvard & Back | Chantal M. Borroum, ESQ. A Dream Is Now Her Reality | Ted Borroum, An Exec Director Morgan Stanley |

Scientist have documented changes that occurred in the bodies of astronauts. Microgravity, or less gravity, led to bone and muscle loss. That remodeling of the bone structure during spaceflight occurs at a rate of about one to two percent per month. Aside from genetics which are the inherited physical characteristics associated with height and weight, creatures adapt to other conditions. Furthermore, adaptations affecting most human beings occurred during a relatively short period of only nine to twenty-five thousand years ago. Prior to that period, humans were more alike than different. The obvious differences seen today were brought about by migration, environmental weather influences, and farm grown products among other things.

One day our students might discover and apply answers to earth's growing problems before it is too late. As requirements continue to climb, one might assume that those accepted by a college will be equal in learning aptitudes and success. But statistics have shown that one person will accomplish much more than another with about the same mental capacity. What is more, the advantages of intelligence may be minimized since clinical tests also suggest that most students, possess far more mental capacity than they will ever use. The important difference, then, is that achievers recognize and work harder to strengthen their innermost talents. Research has also shown that students thrive when teachers hold them to higher standards. It is one thing to encourage and tell a student that they can do better. but their achievements seem to be even greater when they discover for themselves, abilities they were unaware they possessed. Based on my own public-school experiences, I felt the need to give my students an incentive to learn. Therefore, I allowed students who had done well, to engage in challenging activities one day a week for part of the instructional period. Those who fell behind were given makeup work. It didn't take long before students doing makeup work tried harder so that they could rejoin the class in the more challenging tasks that came to be regarded as fun activities. Occasionally, I used a real college level word discovery test. It was an exercise in which students by logical progression solved word association problems. When an administrator, who was visiting, asked my students what they had learned, one student proudly exclaimed, "We learned that we're smart." Various students should be given an opportunity to control the process and ask the questions. These interactions, enable students to lead and better analyze and use their mental gains. Successful self-discovery encourages pupils to attempt more difficult tasks. And while those exercises became enjoyable games for the kids, I marveled at how well they learned because of their self-discovered confidence. We will need minds like these in leading to help save our world.

When I was a principal, I had a fine group of teachers. Some of our core subject teachers included Mr. Kelly and Mrs. Williams—English; Mr. Hall and Mr. Williams—Mathematics; Mr. McBay and Mrs. Powell—Science; and Reverend Nelson and Mr. Johnson—Social Studies. Teachers at the high school receiving new students knew which reading students had been taught by Mrs. Isles. Students entering her room with plans to memorize a series of facts, were subjected to new approaches that broadened their understandings. Those acquired understandings increased their views on issues such as life skills and plans for their personal future. With more confidence in their own decision-making abilities, they devised ways of resolving even more abstract and complex challenges. Another major contributor was Mr. A. Tipps, who was our vocational counselor. Like me, he was a retired Air Force M/ Sgt who displayed in his posture and step the military regimen that had for so long been a part of his life. He rose steadily to direct our district's alternative school program and ultimately became the district's Central Office Vocational Director. In later years he was specifically recruited to serve as the principal of the city of Hutchins' State Jail. The local jail operated under the authority of the Texas Department of Criminal Justice and housed over two thousand prisoners. His instructors focused on programs associated with rehabilitation and the development of technical skills.

Chapter 38
Neighborhood Safety and Social Acceptance

Joyce parked her car in our garage, and I parked my Cadillac on the driveway. One night I was awakened by a noise and as I turned on the lights, I saw a man running away. He had planned to steal the wheels off my Cadillac. Mr. Byrd allowed me to park my Car on his property, but we decided to move to a safer neighborhood. A home auction advertisement caught our eyes. Although we couldn't get to see the inside, we liked what we saw from the outside. The home was built in December 1986 and was sold in January 1987. It was foreclosed on in September 1987, the same year. Most homes in that subdivision were two-story dwellings. Deed restrictions required each home to have two trees, a patio, and a two-car garage. There was only one road into or out of the subdivision which contained a community pool, tennis court, clubhouse, and a children's playground. Before additional homes were added, the southern perimeter of the subdivision was bordered by a forest like array of trees that covered an area which was about three city blocks deep. It was kind of like living in the city with a country-like atmosphere. Behind that forest-like area was a well-maintained public golf course. There was also an elementary school, a tennis court, a playground, a civic association clubhouse, and an area swimming pool.

I owe Alex T, a young neighbor, a debt of gratitude. He was passing by, as I was falling off my ladder. Alex came to my aid and completed the work I had started. Being intelligent and athletic, he received a four-year college scholarship and graduated with honors as a pre-med student. My next-door neighbor, Fred W. helped me with several building projects. He and his wife raised four outstanding kids. Gregory, a graduate of Michigan State is an independent contractor, who owns a company that services million-dollar homes. Katie graduated from Texas A&M University and worked as a state level counsellor in Austin, Texas. Mike and Fred Jr., (AKA Dan) attended Texas A&M and graduated with engineering degrees. I once saw Dan fully dismantle a Ford Mustang down to the nuts and bolts on the family driveway, so I wasn't surprised when he earned a mechanical engineering degree. Hydraulic engineering captured Mikes' attention and he readily displays his competence today as a company executor. I used to tell Fred Sr. that he and his wife should write a book on successfully raising children. When I attended an Ayers family reunion in 2017, I asked Victor and Coleen, to keep an eye on my wife. Their visits and interactions with her exceeded my expectations. During the COVID-19 pandemic they made sure we had a supply of bottled water and other items that were in short supply. This was around the time that I had a conversation with Robert O, while trying to buy scarce items. He and his wife Jackie O, who enjoys her name association with former first lady Jackie Onassis, sets an example for others by regularly walking and interacting with the neighbors. My Army veteran friend and a third Robert on our block, along with his special lady, stopped cars, on our street and handed out PPE equipment. Rick P, a Marine veteran, firefighter, and religious leader, frequently volunteers to help his neighbors. He personally went to Home Depot to pick up and replace both of my faulty smoke alarms. On December 4, 2020, Ben S, my neighbor for thirty years, told me I had a major water leak. When I woke up Saturday morning, Ben, and James S, a fine young neighbor, had found the leak. and called a plumber to complete the work. Vince C in his official capacity, inquired about my chess champion friend Ben Moore. Ben had used me as a reference on a job application. Ben became a lieutenant in the Washington D. C. Capitol Police Department. While most of our neighbors were helpful and friendly, none were friendlier than Lindsey S. During her childhood, her cheerful greetings made the days of those she met a little brighter. Today she is committed to serving others as a health care provider and counselor.

Chapter 39
Driving While Black

I had improved the music system in my 1993 Lincoln Town Car by adding three-way speakers in each front door. I also added an overall system controlling head unit, two 12-inch woofers that were mounted in a sealed trunk enclosure, and a 600-watt amplifier. I then entered an International Auto personal Sound Challenge Association (IASCA) contest in Houston. IASCA promotes electronics through competitions called sound offs. My car took first place in the 600-watt category.

But despite my technical and educational achievements, I had to remember that I was at risk when driving while black or DWB. As I drove through Louisiana in 1994, I saw flashing lights and stopped. I told the officer I was not armed to lessen any fears he might have. At his request I opened my trunk in which I had prominently displayed three documents. They were my honorable discharge, a secondary principal membership certificate, and a Masonic past masters certificate. As he was asking about my unusual display of documents, four more police officers arrived at the scene and parked. Obviously, he had called for backup in anticipation of my resistance. Cars used in a criminal activity could be confiscated in Louisiana. I displayed the documents hoping l would not be viewed as an *unruly Black criminal*. As a Mason, I had established a spiritual bond with God who inspires us to mentally build a better life for **ourselves**. My military papers reflected honorable service to our country, and my principal's certificate indicated I was personally entrusted with the safety of children. Although the police were concerned about my subwoofer whose seal needed to remain intact, they let me demonstrate my system. When the bass became dominant, I had one officer dancing in the street. My display of documents, shows that I am active in many organizations, but my views in this book are strictly my own. I guess the police were satisfied because they allowed me to drive away without a traffic ticket.

Jeremy E Ray American Legion Post #324
On Veterans Memorial Day

I have been active in the boy scouts, the Air Force Sergeants Association, and various educational associations. I am now most active in Post 324 of the American Legion. Wayne Huebner to my left, has been an ardent sponsor of our Boys State events. It is an educational program that encourages top students to gain an understanding, of our constitution and how to run state offices as acting government officials. Boys State was founded in 1935 to counter socialism programs. Our post has produced some excellent contestants and I am honored when Wayne seeks my involvement in this endeavor and the oratorical contests he directs. Other Legionnaires who help on a regular basis include Donna

McCarthy, David Pyke, Curtis Haverty, Bill Novak, Paul Grabert, and others.

Bob Riley is the Legionnaire to my right. One day he handed me a Men of Honor DVD. The contributions of Black military men as combatants were often, diminished. Doris (Dorie) Miller, a Black Navy cook who was killed during World War II. was the first black American to be given the Navy Cross, the second highest combat award after the Medal of Honor. He voluntarily manned the anti-aircraft position of a disabled White seaman during an aircraft attack on his ship. During those days of segregation, men like Doris Miller were assigned non-combat roles. A 1993 study investigating discrimination in awarding of medals, revealed that no medals of honor were awarded to Negroes during WW II. Carl Maxie Brashear, depicted in the Men of Honor movie, was the first Black Navy master diver in 1970, four years after his left leg **was** amputated. Of the 3,470 metals of honor awarded as of June 2015, ninety were awarded to Negro WWII heroes after a reassessment of their deeds.

The flag bearer on the right is David Pyke. He serves as our First Vice Commander. During a joint meeting and festive, family celebration, I decided to attend only the meeting portion of the get together and then return to listen as my wife sang in church. The Legion meeting was already crowded with over 100 people. Dave who assumed the role of an usher that day, got a chair and squeezed me into a space besides his family. He then resumed his duties as an usher. I really value the fact that he trusted and respected me enough to place me amid his loved ones.

Our Post Commander, Stefanie Otto was an anti-vaccinator. But she allowed herself to be vaccinated to protect our members. In addition to being a wife and working mother, she also exhibits agape love for all of us. Our talented adjutant, Larry Dick, routinely redistributes timely, and accurate data to keep our members well informed. I wonder how he handles all the tasks. Ron Lance, our long-time treasurer performed his duties admirably despite health issues. Past Commander Jimmy Baughman has, on occasions filled many positions. He is a big man who readily fits the description of what one might expect a Marine to be. Yet, when I teasingly challenged him one day with the words "You can't hurt me", his calm and measured reply was, "I don't want to." He knows who and what he is and does not feel the need to prove it. I can't visualize what the financial status of our post would be without Curtis Haverty's annual fund-raising 5K runs. Trenton Davis, a Marine, was the architect of a highly successful fund-raising drive in 2020. Other hard-working performers and achievers include John Link, Charles (Trey) Pryor, and Bill Brown, a driving force capable of handling multiple Post 324 positions.

Our American Legion Post took part in the dedication of a very much renovated Fire Station just across the road from the Veterans Memorial Cemetery. The ceremony was held on May 30, 2022, which was Veterans Memorial Day. We were there to display the American and Texas flags as well as six additional flags representing our military services. I almost ruined the entire ceremony because I was thinking about a joke I had told earlier regarding two men in kilts. I said to the men, "Don't hit me but I really don't want to see what's under those skirts." As a result of being distracted, I didn't get a full understanding of how to secure the rope I used to raise and lower my POW flag. The poles on which I had hoisted flags in the past had hooks to tie the rope around. Fortunately, Mike LaRosa, the legionnaire next to me, realized I was struggling. Mike had begun his career in the Coast Guard as an enlisted man before being promoted to the rank of Warrant Officer 4. Ultimately, he was commissioned as an officer. If he had not come to my aid, I would have failed to display the POW flag properly. The POW flag reminded me of an earlier Legionnaire gathering of over 200 attendees about thirty miles south of Houston several years ago. The dinner and ceremony recognized two U. S. Army men who had been captured and held as POWs in North Korea. Joyce and I were seated at the same table as the former POWs. We were therefore able to meet some government officials such as Lieutenant Governor Dan Patrick and Congressman Al Green.

Me, Joyce & Congressman Al Green

Ceremonial Kilts

Our Briefing

Honored To Be Part Of The Ceremony

Mission Accomplished

Chapter 41
My Issues With Luck

No matter when in life, we as individuals decide to make a change, it is possible to do so. Early in life my only desire was to get off welfare. I may not have known how I was going to do that, but you could say I was *lucky*. Luck for me was two neighborhood adult brothers, who effectively kept pedophiles from continuing to prey on kids. Their actions allowed me to escape from a despicable situation while I was still mentally able to regain a measure of my self-respect. Luck was a tall Italian who delivered me to a parent instead of the police. That decision kept me, innocent or guilty, from having a criminal record. If I had **failed** the background check performed by the Office of Special Investigation (OSI) I would have been ineligible to work with classified information. Luck was an intelligent H. J., who, unknowingly, inspired me to study and perform academically so that my teachers could get a glimpse of what I could possibly become. Luck let me work in a career field where I collected important information and then briefed aircrew members who unselfishly shared their trained and often superior insights with me. If my original job assigner had realized I was Black, I would not have been selected to attend a then still segregated operational intelligence school. Les Brown, an inspirational speaker said if you find yourself in a gathering where you are the smartest one in the group, it's time to find another group. I feel fortunate not to have been influenced by such an opinion and humble enough to remain in the group with Airman Bussey. His drive led me to believe I could be the deciding authority in how to meet the educational needs of children. Teaching was a career that was made for me.

My mother tried to instill in her children a strong love of God. But it is one thing to declare oneself a Christian, and then engage in actions not approved by God. As a submarine Christian, I generally surfaced from questionable desires on Sundays and then sank back into my world of temptations the rest of the week. But God watches us throughout the week. I am so grateful that He had a goal for me. I now really understand that it was God and *not luck* that guided me. He let me chart a positive course so I could effectively meet the needs of His children.

Whitney Houston, an internationally recognized performer, with a once in a lifetime singing voice, said, "Crack is whack." as if the higher priced drugs she could afford, made her superior. But another long-addicted drug user and father of a Heisman Trophy winning football running back, was quoted as saying, "There are no successful drug users." I feel inclined to agree with him. Drugs and alcohol have ruined the lives of too many people that I care about. My hero, Tommy Lee, received a dishonorable discharge and became a drunken derelict in our hometown before his death. But he was and always will be a hero who helped me. I would like to have been there to support him when he was so much in need of help. I hope that each of us will pursue the drug free career that fulfills our innermost desires, hopes, and dreams. Some of us, as Lincoln stated, must create our destiny and that won't be easy to do. We must then record our own progress to make sure we stay motivated and on track. Don't expect others to make choices for you; you must do that for yourself. I made a responsible decision at a 1949 Billy Eckstine concert in Trenton, New Jersey. Billy was an American pop singer and a bandleader during the swing era, who was noted for his rich, almost operatic bass-baritone voice. During a restroom break, I watched two guys swapping a funny shaped cigarette. They offered me a puff on it., but I said no. I later discovered it was a marijuana cigarette. I avoided a practice whereby I most likely would have been hanging around users of multiple drugs and become one of them. Unfortunately, I didn't avoid being around cigarette smokers in basic training. I learned to smoke cigarettes during breaks in our daily military training. Those breaks were always followed by our flight marcher's commands ordering us to show him *rear holes and elbows* as we bent over to pick up discarded cigarette butts. It took me eleven years to stop smoking. Drugs, according to research, can physically alter the brain to a point where addicts lose control. In addition, drug dealers often alter their products to make them more addictive to unsuspecting customers. It's a lesson learned only at the point of dying by too many drug addicts.

Chapter 42
Advice

I usually avoid giving career advice, but when students insist, I suggest they make a list of the things they enjoy doing. They should then study each listed activity in-depth to find out if any of those interest, along with formal training, could lead to an economically and emotionally satisfying livelihood. In situations where their goals require the study of disliked topics, they should understand that whatever it is that they are required to learn should be learned well. The awarding of scholarships is based on all-around excellence. Additionally, knowledge derived from less favored fields, may at some point help them do their preferred jobs better. Those seeking financial assistance may apply for assistant programs such as FAFSA, which is the Free Application for Federal Student Aid. The form is currently online at fafsa.ed.gov and it is a recommended link to the form. In my own case, I had no ideas about what I wanted to do. Discovering what it is in life that we really desire to do takes time, but the benefits of such discovery and then doing the type of work that one really enjoys, is immeasurable.

I tried to keep children from giving me gifts. But the pain of anticipating that her gift would be rejected, was, so apparent in Jaime's eyes that I accepted her gift. She would be pleased to know, that I am attempting to use her gift to make a point in my book.

Jamie

See, Speak, Hear No Evil

The visual messages portrayed by the three little pigs discourages evil as it is seen, spoken, and heard. But many of our leaders have interpreted that to mean you should close your eyes to block from your view the pain, suffering, and terror that mob justice or authority delivered brutality may inflict. Others remain silent when witnessing a crime. Do people really believe that by just covering their eyes and ears that they won't be able to imagine the cries of 14-year-old Emmett Till as he was being lynched? His accuser later confessed that she had lied in court. On April 03, 2022, President Biden signed the Emmett Till Anti-Lynching act into law. Yet, the hue and cry of anti-Critical Race Theory opponents gets louder to keep children from learning about such racial horrors in school.

People often act on what they want to believe rather than the facts. But never has hate and violence been practiced more openly, even within families, as it is today. Who do we really help if our covered eyes can't see criminal acts? Also, you don't have to physically touch the rope to be complicit in a lynching. How can we be against abortions on the one hand and support policies that allow children to be separated from their parents, as a tool to discourage legitimate asylum seekers? How can we justify the widespread mislabeling of all asylum seekers as drug dealers, murderers, thieves, and rapists to deliberately poison the minds of our citizens against desperate families? Some political figures have made efforts to alter **or** restate the inspiring message contained in a sonnet written in 1883 that is now displayed on a plaque placed in 1903 at the base of the Statue of Liberty. Originally, the statue was to be placed near the Suez Canal, but on July 4, 1884, France donated the Statue of Liberty without a base or a pedestal to the United States in a gesture of friendship.

The sonnet, or 14-line poem, was written by Emma Lazarus as part of an American fund-raising effort to pay for the base and pedestal. But only nine days after being inaugurated, President Trump gave Americans a real view of his less than caring regard for the downtrodden. Kenneth T. Cuccinelli II, a top Trump immigration official, restated the poem's message to mean: *Give me your tired and your poor who can stand on their own two feet.* Just another way of saying don't come here if you are going to burden us with your needs. He also tried to convey the idea that the huddled masses referred to in the sonnet should be skilled and talented Europeans. The sonnet's actual wording declares, *From her beacon-hand glows world-wide welcome.* Laws should be enforced rather than just manipulated.

Dr. Martin Luther King believed in change without violence. Malcom X disagreed. The Pilgrims landed safely on Plymouth Rock after a voyage of almost two months. Malcolm X said that as slaves we didn't land on Plymouth Rock. Plymouth Rock landed on us. He also believed it was criminal to instruct a man not to defend himself when he is the constant victim of brutal attacks.

I believe in my Christian God, but I am also aware that about two thirds of the earth's population worship other gods or no god at all. Still, I am baffled when people, who identify themselves as Christians, knowingly violate the first of God's ten commandments. How much clearer could God be when He says *you shall have no other gods before Me* and forbids giving the glory and honor to any creature which are due to God only? Yet, we find political, church, business, and activist leaders urging their followers to support the dictatorial actions of mortals who routinely use hateful rhetoric, division, lies, or Malcom X's by any means necessary methods to achieve their self-serving desires. Why don't we hear from so many of our elected leaders as they complicitly, through their silence support distorters of the truth? If this world, filled with mistrust doesn't reverse itself, then we will all succumb to the overriding issues of Vladimir Putin**'s** nuclear war threatened expansion type aims, as well as the many other issues threatening the continued existence of life on our planet.

Chapter 43
Resistance To Change

No matter how others may choose to classify me, all I ask is that people be allowed to develop economically, spiritually, educationally, and ethically without discriminatory limits. Scientific studies reveal that mankind's adaptation to environmental conditions and his gradually increasing consumption of farm grown products are factors contributing to the differences in man's color and appearance. Nearly all geneticists (students of the variation of inherited characteristics), scientists and biologists acknowledge that there is no biological and social scientific foundation for race. If all human beings were to live in the same environment, and had similar diets, then racial groups theoretically would be indistinguishable from each other. Unfortunately, too many people limit their acceptance of information to only the things they want to believe. Even worse, they try to force their unsupported by fact beliefs on others.

It is difficult to unite people, who hate and often do not agree even with themselves. For example, Senator Mitch McConnel initially suggested President Trump's impeachment trial should be delayed until Trump had been given time to prepare a defense. Later he voted to acquit Trump because the delayed impeachment trial was now unconstitutional since Trump was no longer in office. Following his own acquittal vote, he then delivered a blistering attack against Trump's January 6 mob inciting attack on Washington, D. C. But he then reversed himself again and influenced his followers to reject any hearings about the January 6 insurrection that might uncover his party's involvement. The senator is obviously playing both sides against the middle with his obvious long-range goal being to regain control of the senate after Trump's base of support is reduced. However, Trump's blistering responses to the senators shifting positions may undermine McConnel's future position and influence.

Matthew Gaetz, a representative for Florida's First District is a fiery Trump supporter. Gaetz had questioned Secretary of Defense Austin about the dismissal of Space Force Lt. Colonel Matthew Lohmeier in May of 2021 for his partisan political activity. Lohmeier had insisted that military diversity training was rooted in *critical race theory and Marxism*. Because of the still ongoing investigation, regarding the matter, Defense Secretary Austin limited his response. However, General Mark Milley, the Chairman of the Joint Chiefs of Staff, said it is important for those in uniform to be open minded, nonpartisan and well read. and even though he had read the teachings of Mao Tse Tung, Karl Marx, and Lennon, just being better informed about them didn't make him a communist. The general also wanted to understand White rage. He let it be known he was offended by the thousands of Americans who were inspired to assault our government, police force, and elected officials. In a new book, "I Alone Can Fix It" by Washington Post reporters Carol Leonnig and Phillip Rucker, the general compared Ex-President Trump to Adolf Hitler while alluding to the coup preceding Hitler's rise to power in Germany. Milley stressed that his oath was to uphold the Constitution and he would have resigned rather than enforce unlawful commands. And speaking of military matters, on a much broader scale, there is no way we can ignore the potential threat of worldwide nuclear annihilation openly referred to by the Russian dictator Putin. Once again, he is acting on a plan to regain Russian control of territory and worldwide influence. Like the dictators before him, he is suppressing the truth and misleading the Russian people. Chemical and nuclear war are an ever-present threat if men like Putin remain in power. As a history major, I tend to dwell on the past and its potential effects upon the future. When we teach history, its truths must be accurate and above all else, it must not contain selective omissions.

Chapter 44
Educational Structure

My early preparation to become a teacher—unknown to me at the time—began when I was mistakenly chosen to receive intelligence training in classes that weren't integrated. In 1825, articles written by James C. Carter appeared in a Boston newspaper. His articles suggested that knowledge attained through research would yield information about future successful teaching. However free public education began in Boston schools in 1820. In 1827, Massachusetts passed a law making all grades of a public school open to pupils free of charge. Compulsory education, however, was not required until 1852. Today free public education is offered for twelve to fourteen years, on a state-by-state basis. Twenty states currently offer a version of tuition-free community college type grants for students willing to pursue skilled job training. The training is generally in fields such as commercial truck driving, welding, computer proficiency, nursing, and other widely needed job skills.

I personally encountered parents trying to control the information received by their children. On one occasion, a parent who was a school board member, expressed a concern that I was demanding too much of her daughter. In that case Phyllis' achievements exceeded even my expectations. But, since widespread agreement even within groups does not exist, it is up to professional educators, who also have their own differing opinions, to meet, discuss and peacefully resolve those differences. The mutually agreed upon curriculum or the content of subjects that students will be exposed to, as well as the grade level is then adopted. Parent Teacher Associations and parents in general are provided opportunities to recommend changes at school, district level, and even state level board meetings. The process ensures that uniform and agreed upon topics will be taught throughout each city and state.

I believe that what is taught, to whom it is taught, when it is taught, and how thoroughly it is taught are important issues in education. In my answer to how and at what rate should children grow and develop, I'll attempt to answer by comparing the teaching of mathematics with the teaching of history. Children begin learning math starting with the counting of whole numbers such as zero through nine and beyond. The four functions of addition, subtraction, multiplication, and division as well as the order in which such problems are to be solved comes next. As the child progresses, we move on to beginning algebra, trigonometry, geometry, calculus, etc. Neil deGrasse Tyson is a well-known science communicator and American *astrophysicist*. He is now the Hayden Planetarium's Frederick P. Rose director at the Rose Center for Earth and Space. He is also one of the research associates of the American Museum of Natural History's department of astrophysics. Physics is a natural science that involves the study of matter and its motion through space and time, along with related concepts such as energy and force. Neil's first of six books on astronomy was published in 1988. Those books educate individuals on topics like star formations, exploding stars, dwarf galaxies, and the Milky Way's structure. But there are other astrologers whose achievements required a thorough and critical understanding of math. For example, the James Webb Space Telescope, a ten-billion-dollar NASA investment, was launched in 2021 on Christmas day. It was designed to enable man to obtain a view of the origin of the universe by gathering infrared light from galaxies that are billions of light years away. Webb showed his genius in various posts from 1961 to 1968 and is said to have contributed significantly to America's moon landings.

Words That Wound: Critical Race Theory, Assaultive Speech, And the First Amendment, is a book by Mari J. Matsuda, Charles R. Lawrence III, Richard Delgado, and Kimberle Crenshaw. This is the first book to gather the key essays on CRT and it allows a person to have a generalized idea of what CRT was originally about. But the book also views CRT problems as being unsolvable in that social reform requires that everything must change at once. Otherwise, the partial changes are swallowed by the unchanged elements so that we remain roughly as we were before. Critical race theorists adopt a stance that presumes that racism has contributed to group advantage or disadvantage along social lines. The disparity in the numbers indicate that those differences are in income, imprisonment, health, housing, education, political representation, and military service. Power structures that benefit one race over others are inherently racist. Anyone who is content to live within such a structure and not be actively against it is, for all practical purposes, supporting doctrines that are not applied equally.

We have been made aware of the danger associated with the use of fossil fuels and how global warming is causing a reduction of our artic ice deposits in areas like Greenland. But the gradual deforestation in tropical rain forests is also of concern. This is because photosynthesis, which is a process used by plants and other organisms to convert light energy into chemical energy, through cellular respiration, can later be released to fuel life forms. *Trees, plants, and some bacteria interact with water during the photosynthesis process that develops the oxygen we breathe.* We must be concerned with tree removal from land that is then converted to non-forest use. The process converts forest land into farms, ranches, and urban use. The most concentrated deforestation occurs in tropical rainforests. About 31% of Earth's land surface is covered by forests at present. But that is still one-third less than the forest coverage before the expansion of agriculture thousands of years ago. Half of that loss has occurred in the last century. About 58,000 to 70,000 square miles of forest, an area the size of Belgium in Europe, are destroyed annually. About 2,400 trees are cut down each minute in rain forest. Trees and plants pull carbon dioxide from the air and turns it into sugar through photosynthesis, while releasing oxygen. This sugar is used to build organic matter. I now know about the existence of bacteria in forms other than the one that almost killed me at age three. But concerns about bacteria are not to be dismissed frivolously because NASA scientists are concerned about "thermokarsts" or the lakes that appear as the permafrost in Alaska thaws. When thawed, these lakes are sending bacteria developing methane into the atmosphere. The lakes are so full of bacterial climate-damaging gas that it can be seen bubbling to the surface. More and more of these lakes are appearing as Alaska's permafrost thaws with rising temperatures and increasing forest fires, according to a 2021 study.

Many races throughout the world have been victims of racial injustice. But whether the numbers are large as in the holocaust or small there is great suffering. For example, *La Matanza* reflects a period when Mexicans were routinely slaughtered. American settlers, reportedly supported by law enforcement officials, shot, and lynched more than 200 people. The people in those 1915 incidents were labeled as criminals to justify their slaughter as well as the confiscation and seizure of their property in disputed Rio Grande Valley areas. Anti-immigration laws continue such unjust practices today.

Many other ethnic groups have also felt the pain of racial injustice. In our world otherwise upstanding people have turned a blind eye to such misery. If you thought the Emancipation Proclamation ended slavery, think again. Peonage for example is a practice where people are forced to pay off their debts with work. Although outlawed by Congress in 1867, more than eight hundred thousand (800,000) Black men in some southern states were forced into peonage by questionable methods. It is estimated that forty percent of the men or about three hundred and twenty thousand died because of the harsh conditions in which they were forced to work and live. Peonage, which in essence, was a return to slavery, was not completely eradicated until the 1940s; about eighty years after the Emancipation Proclamation.

Carol Anderson is the Charles Howard Candler *professor of African American Studies* at Emory University. Her research focuses on public policy about race, justice, and equality. According to her and others, CRT studies were conducted in the 1970s to try and understand why the U S civil rights movement was not only losing its momentum but was in danger of being reversed. The study pinpointed systemic aspects of our laws that tend to maintain race-based oppression and White privilege. The term critical race theory was created by Kimberle Crenshaw. In the 1970s and 80s she was one of the pioneers connected to a Harvard Law School group led by Derrick Bell. Kimberle described CRT as neither Marxist nor racist, but as a practice or a way of seeing how the fiction of race has been transformed into concrete racial inequities. White power supporters have opposed efforts to expose their children to race based CRT instruction. I don't remember hearing any complaints when Black children had to read *Little Black Sambo* with its self-image demeaning content. We have avoided exposing our children to true historical facts that may or may not affect what they do at some future date. We should instead gradually teach disturbing things within the limits of a child's ability to learn **and** understand, accompanied by a realization that they are not responsible for the past actions of other people. *I wonder what PhD Neil DeGrasse Tyson would be doing today if he had not been permitted to study calculus or physics.* The simple answer is something else. Years ago, we heard the cry from right wing Americans that integration is communism. During the many Black Lives Matter movements shown on TV, it appears that more Whites are taking part in Black Lives Matter rallies. White supremist hope to slow that trend by keeping their children uninformed about still ongoing terrible issues?

I view CRT, by any name, as a real and important part of history that should be taught. Plato, an Athenian philosopher, and founder of the first institution of higher learning in the western world, gave this advice to parents. Don't force your children into your ways for they were created for a time that is different from your own. Teaching CRT should not be regarded as promoting the hatred of white people. If done properly, there will be fewer conflicting version of history. Historical revelations will not be based on variations, cover-ups, or omissions of the truth. Hidden events allow children to remain ignorant about still ongoing horrors and they will have little empathy for those who may still be suffering. Those demanding that White children not be exposed to CRT lessons are, in essence, demanding that such children be allowed to remain indifferent to the suffering of others. Please do not keep our children from learning about past horrors to the point where they will be unable to recognize reemerging signs and unwittingly become a part of developing trends in which horrors are repeated. How many men must be sacrificed in peaceful civil activities for widespread inequities, favoring one or more races, to end?

Chapter 45
Ideological Loyalty

Ideological loyalty is the practice of being loyal to a particular belief, even in the face of evidence which clearly shows it to be an inaccurate model of reality. It is the cause of ideological protectionism, and is often a powerful tool for causing large numbers of people to take actions that are generally against their own best interests while believing they are doing so for a greater good. Evidence that points to inaccuracies is suppressed rather than corrected. Disagreements with the basic tenets is likely to result in a person's removal. It is permissible for those supporting the basic tenets, whether right or wrong, to use extreme anger, violence or even cause death to further their aims. This current situation in America seems to be a summation of what Mr. Amos, the cult follower who I listened to when I was just three, kept saying. That of course is, "There is no wrong way to do the right thing."

In 1970, I directed *Ventura*, a sidewalk artist, to paint a portrait for me. I intended to use the portrait to warn people of color to be wary of distractions such as flowers, that might keep them from realizing they were drifting into danger. That portrait has significance today for people of every nationality. The black silhouettes, in the absence of light, have no color or identifiable features. Please consider from a purely scientific point of view, the factors that influence how Homo Sapiens look. Each silhouette now could represent three much broader categories; (1) the **worldwide united** leadership of all nations; (2) the varied and, for the most part, united worldwide populations, and (3) the peacefully negotiated resolution of acceptable viewpoints. It is only by joining together as if we were a single society and combining our efforts, that we may yet avoid our impending planetary disaster.

VENTURA 1970

I remember Vice President Al Gore addressing the issue of global warming more than two decades ago. I have also taken note of the international cooperative efforts regarding Global Warming that have been increasing since 2015. The United States under President Biden has, to some extent, resumed its international efforts to address the crisis. The efforts by multi nations pledging money and other resources to address the problem are promising. But failure is not an option because nature sets its own course based on ongoing developments and has no favorites. According to scientists the four and a half billion years that nature has taken to establish suitable conditions for living creatures, could be reversed beyond recovery in a matter of decades. As I stated before, no person, group, nation, or force managed by man, has *ever* controlled nature and **its** invincible powers. The greatest obstacle standing in the way of reversing this impending disaster, is disunity and hate among the citizens of the world. The industrial ambitions of some countries as well as the Invasion of countries like Ukraine, do little to promote a reduction in the use of fossil resources or avoidance of potential nuclear disasters. Any hope that mankind will act in unison for the common good, is very low on any measurable scale of statistical probability. According to a September 2022 United Nations report, weather disasters costing $200 million a day and the irreversible climate catastrophes looming show the world is heading in the wrong direction. Even if the world somehow manages to limit future warming to the strictest international temperature goal, at least four, out of the sixteen currently identified Earth-changing climate tipping points are still likely to be triggered with others to follow.

What appears to exist in our world's present-day climate of opposing views is excessive *hate*! How can we persuade the masses and leaders, with long established self-serving aims, to alter their priorities and join realistically rather than symbolically, in the effort to stabilize the canoe in my Ventura portrait that keeps our heads above water? How can we encourage the massive groups of people who constantly disagree, to pick up an oar and row in unison with others? We need to be united in our efforts to *reach that symbolic glimmer of light that is barely visible above the* canoe. That light will become brighter as we increase our efforts and approach safety together. It took over two million years of human prehistory and history for the human population to reach the one billion mark and, according to scientists that number has increased to 7.9 billion as of November 2021.

The James Webb Space Telescope has made some impressive discoveries such as the detection of carbon dioxide on an exoplanet or planet outside our solar system. However, scientists with Harvard and MIT are warning that James Webb's data could be lying to astronomers. They say that the available model used to determine how photos pass through a material isn't up to par with the quality of data delivered by the James Webb telescope.

Excuse me for using such an old expression, but it is time that all these naysayers get their heads out of the sand and aid in the effort to save not only our planet but themselves as well. Even if the James Webb telescope or other space explorations lead to the discovery of a planet that we can live on, how would we go about transporting the number of human beings on earth and the products needed to sustain living creatures on any planet? What we can't afford to do is take two steps back because of the expansionist desires in land and industry on the part of some nations, who even if they win, would do so only momentarily. If the world doesn't unite and change its priorities, we will all become uncharted drifters, floating aimlessly into the dangers represented up high by the partially hidden snake in the tree, or *by the surface* dangers symbolized by that massive barely visible *paw* in the bushes; and finally, by the unimaginable hazards lurking below our dying coral reefs and the vast depths of our polluted oceans and rapidly depleted land bound waterways.

During periods of drought in the 1940's, scientists turned to airplanes and cloud seeding with sand, dust, and dry ice to make it rain. This method used existing clouds, rather than the earlier attempts to create clouds. In today's world, for the sake of making money, mankind, with the blessing of our Supreme Court, has allowed proven global warming forces such as our industrial power plants to defy EPA guidelines and continue operating. Theoretically, once man-influenced global warming has reached what appears to be an overall destructive point, unlike the dissipated clouds which no longer contain any rainwater, it will continue its global destroying development and cannot be stopped.

I'm no longer that youth who in the past, without any noteworthy achievements, deluded himself into believing that things would continue to be okay. Now this book is part of my effort to encourage all of us to get involved to eliminate hate and conflict. Realistically there is very little possibility that things will get better. However, whether viewed as naïve or not, I won't quit and leave it to others, including some I may have taught, to solve earth's problems. I regret passing my *middle school Sarah's lust problems to be solved by others.* Now, I'll continue to be involved in working to solve earth's problem for as long as I live. The world is facing a near impossible series of challenges. But the solution to every impossible, but hopefully solvable problem must begin somewhere. So, no matter what our religious affiliation may or may not be, or what we want to believe, we must all find a way to rearrange our priorities, come together, and devote all our efforts to saving our planet first.

Owning my home suggest that I have attained a goal that many desire. But since I began to focus less on the image in my mirror as well as my self-centered needs, my outlook has broadened. Now, with my ever-expanding understandings, I can see others who despite doing everything they can, have still fallen short of meeting their essential needs. I want to keep evolving into a more caring individual who is trying to make the world a better place for all of us. It isn't what I do for myself now, but what I am willing to do for and with others that makes **my** existence on this earth meaningful. We, the people of the world must replace world *hate* with global unity, hard work, scientific discoveries, preservation of resources, recycling, adequate financing, and never last or ever the least, LOVE!

My student Nora B is celebrating her quinceanera. It is a ceremony when a Latina girl turns 15. Nora's world always contained sincere expressions of agape love that contributed to a world of tranquility and peace for those in her vicinity. Tameka and Phillip's children, whether awake or sleeping, also demonstrate the gift of agape love. It would be wonderful if the love expressed by these youngsters could exist throughout the world. I can't think a better time than right now for all of us to join in the promotion of Genuine *LOVE* for all we meet and greet. For some of us that will be a very important beginning step toward doing what we can to unite and work actively with many others in our efforts to save our planet.

Kai and Tyler

Devin and Tyler